3 in 1
Practice Book

Practice
Reteach
Spiral Review

Grade 2

Visit *The Learning Site!*
www.harcourtschool.com

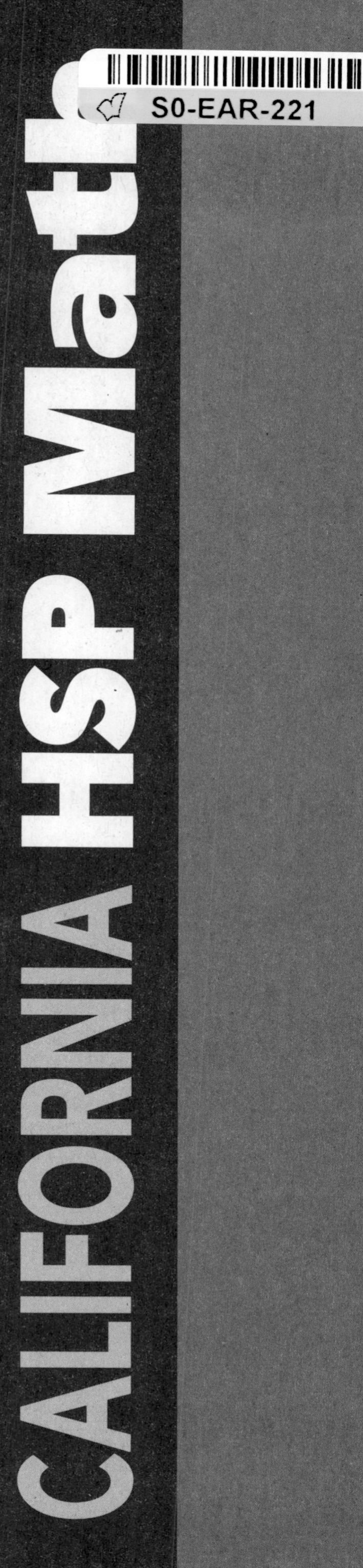

Printed in the United States of America

ISBN 13: 978-0-15-383383-0
ISBN 10: 0-15-383383-1

2 3 4 5 6 7 8 9 10 0956 17 16 15 14 13 12 11 10 09

Contents

Key: PW Practice Workbook RW Reteach Workbook SR Spiral Review

Contents

Key: PW Practice Workbook RW Reteach Workbook SR Spiral Review

Contents

Key: PW Practice Workbook RW Reteach Workbook SR Spiral Review

Contents

Key: PW Practice Workbook RW Reteach Workbook SR Spiral Review

Spiral Review

Write the sum or difference.
Write the numbers in the fact family.

1. $8 + 7 =$ ______ $15 - 7 =$ ______

$7 + 8 =$ ______ $15 - 8 =$ ______

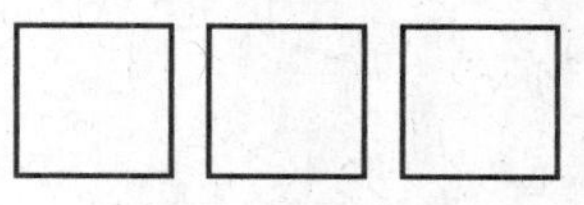

Circle the pattern unit.

2.  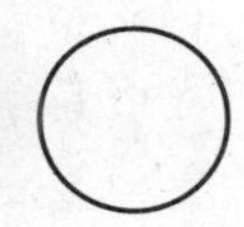 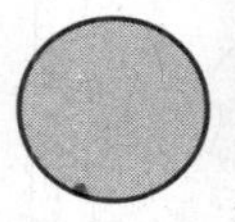  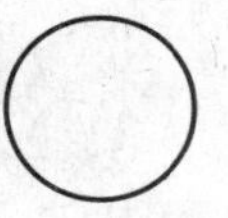 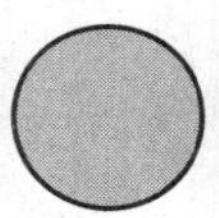  

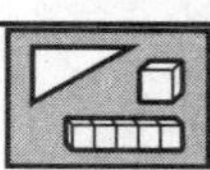

Use a clock.
Show the time.
Write the time.

3.

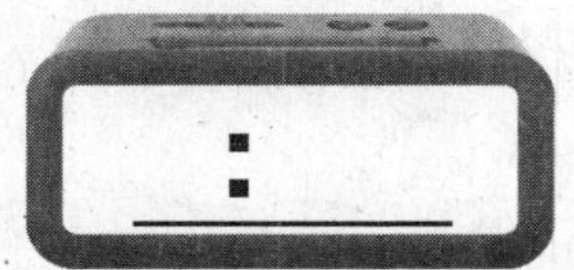

4.

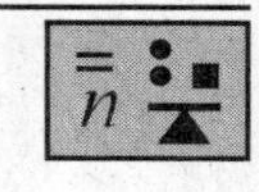

Circle **add** or **subtract**.
Write the number sentence.

5. Jack has 12 bananas.
He eats 4 of them.
How many are there now?

______ bananas

add subtract

______ ◯ ______ ◯ ______

6. Alice has 6 cherries.
Todd brings 7 more.
How many are there now?

______ cherries

add subtract

______ ◯ ______ ◯ ______

Name ______________________ **Week 2**

Spiral Review

Follow the rule.
Complete the table.

1. Rule: Add 6.

In	Out
2	
3	
4	
5	

Write the sum.

2. $5 + 3 + 7 = $ ____

3. $2 + 6 + 9 = $ ____

4. $2 + 7 + 8 = $ ____

5. $3 + 4 + 7 = $ ____

Write A, B, or C to show which group each figure belongs.

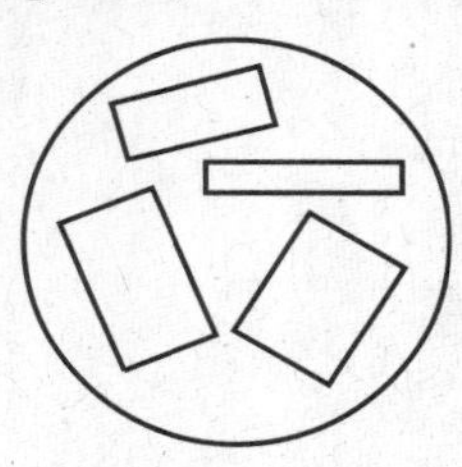

A

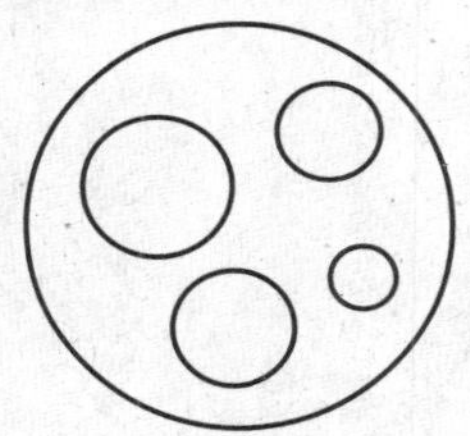

B

C

6. 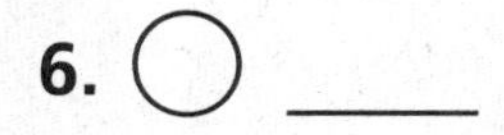

7. ____

8. ____

Circle the object you could trace to make each figure.

9. 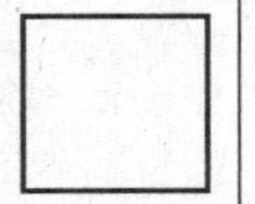  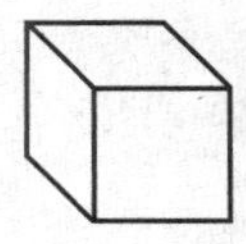  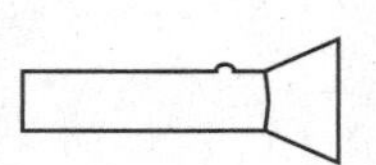

10. 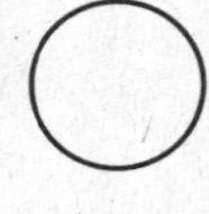 

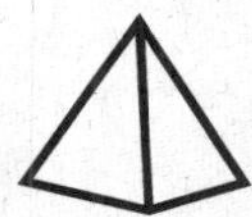  

Name ____________________

Lesson 2.1

Understand Subtraction

You **subtract** when you **take away** part of a group.
You subtract to answer the question "How many are left?"

Kat has
8 stickers.

She gives
2 away.

Kat has 6 stickers left.

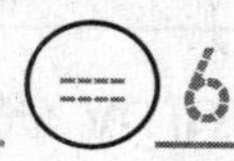

8 − 2 = 6

You also subtract when you **compare** two groups.
You subtract to answer the question
"How many more?" or "How many fewer?"

Lon has
5 stickers.

Ann has
3 stickers.

Lon has 2 more
stickers than Ann.

5 − 3 = 2

Draw a picture of the problem.
Write the number sentence. Solve.

1. Troy has 6 pencils. Two pencils roll off his desk. How many pencils are left?

pencil

____ ◯ ____ ◯ ____

____ pencils

2. Ari has 6 pens. Jim has 3 pens. How many more pens does Ari have than Jim?

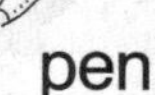

pen

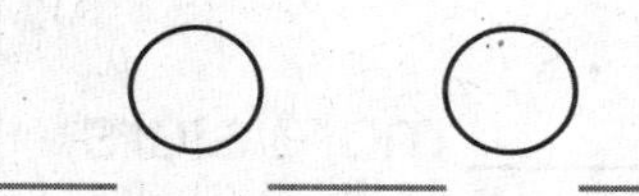

____ ◯ ____ ◯ ____

____ more pens

NS2.2 Find the sum or difference of two whole numbers up to three digits long.
AF1.3 Solve addition and subtraction problems by using data from simple charts, picture graphs, and number sentences.

Name ______________________________

Lesson 2.1

Understand Subtraction

Write the number sentence. Solve.

1. There are 8 kites in the air. Then 3 kites fall to the ground. How many kites are still in the air? KITE	8 ◯ 3 ◯ 5 kites
2. Tammy eats 5 nuts. Jon eats 7 nuts. How many fewer nuts does Tammy eat than Jon? nut	____ ◯ ____ ◯ ____ fewer nuts
3. LaBron sees 8 birds. Bret sees 2 birds. How many more birds does LaBron see than Bret? bird	____ ◯ ____ ◯ ____ more birds
4. There are 9 towels at the pool. Bella takes 1 towel. How many towels are left? towel	____ ◯ ____ ◯ ____ towels

Problem Solving

Use the picture to complete the sentence.

5. There are _____ more stars than bells.

Spiral Review

Complete the fact families.

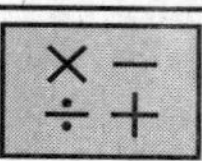

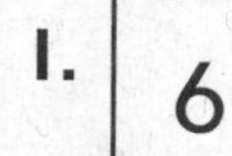

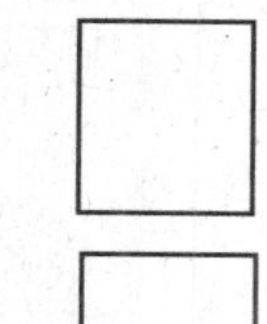

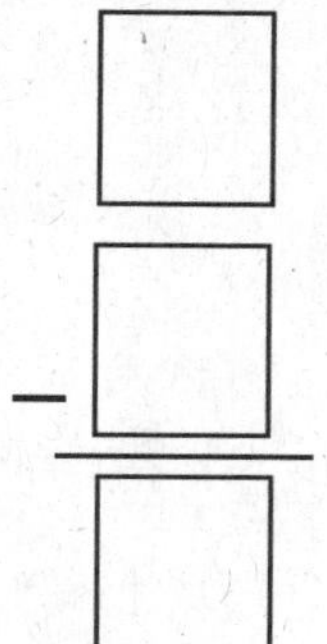

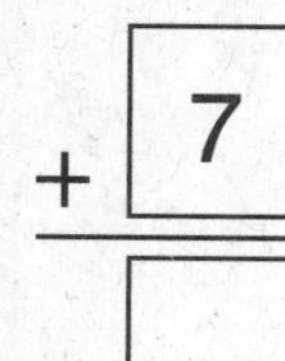

1.

6	☐	☐	☐
+ 7	+ ☐	− ☐	− ☐
☐	☐	☐	☐

Sports We Like		Total
Basketball	𝍸	5
Football	𝍸 ||	7
Baseball	|||	3

Use the tally chart to answer the questions.

2. How many children chose football?

_______ children

3. Which sport did the fewest children choose?

____

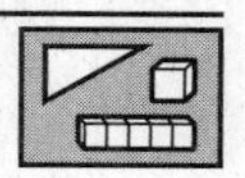

Use the real object and ⊂⊃. Estimate. Then measure.

Object	Estimate	Measurement
4.	about _______ ⊂⊃	about _______ ⊂⊃

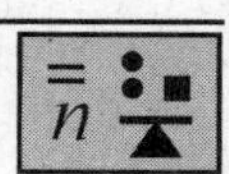

Count back to find the difference.

5. $\begin{array}{r} 5 \\ -\ 2 \\ \hline \end{array}$

6. $\begin{array}{r} 9 \\ -\ 1 \\ \hline \end{array}$

7. $\begin{array}{r} 7 \\ -\ 2 \\ \hline \end{array}$

8. $\begin{array}{r} 6 \\ -\ 3 \\ \hline \end{array}$

Spiral Review

Make groups of ten.
Use Workmat 3 and 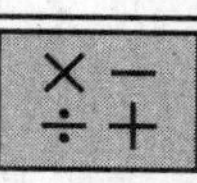.
Write how many tens and ones. Then write the number.

1. 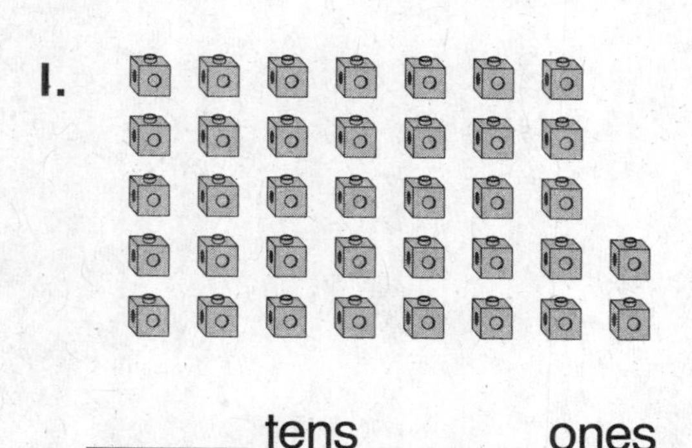

_____ tens _____ ones

_____

Use plane figures.
Make a tally chart to solve.

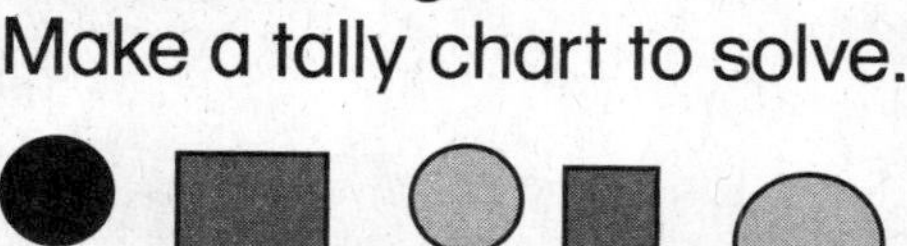

Plane Figures		Total
■	||	2
▲		
●		

2. How many more circles are there than squares?

_____ more

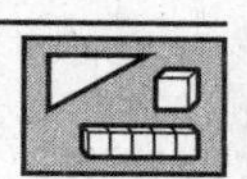

Use the solid. Write the number of flat surfaces and corners.

3. A cube has _____ flat surfaces.

4. A cube has _____ corners.

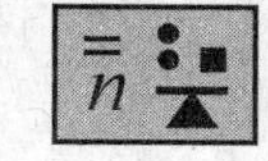

Decide which operation to use.
Then write the number sentence. Solve.

5. Jeremy has 8 crayons.
He gives 3 crayons to Natalie.
How many crayons does Jeremy have now?

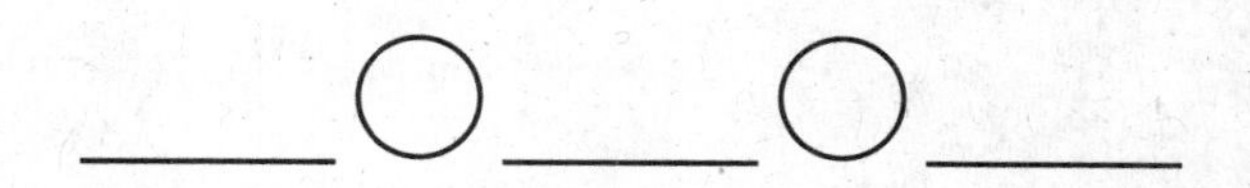

_____ crayons

Name______________________________________

Lesson 3.2

Tens and Ones

You can use a model to find how many tens,
how many ones, and how many in all.
First, count the tens. Then count the ones that are left over.

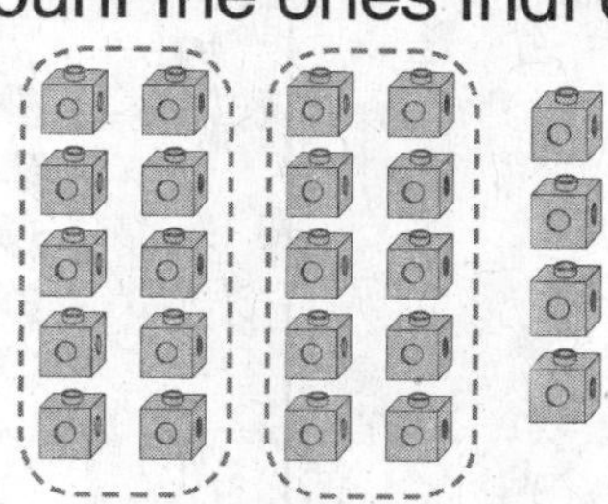

There are _2_ tens. There are _4_ ones left over.

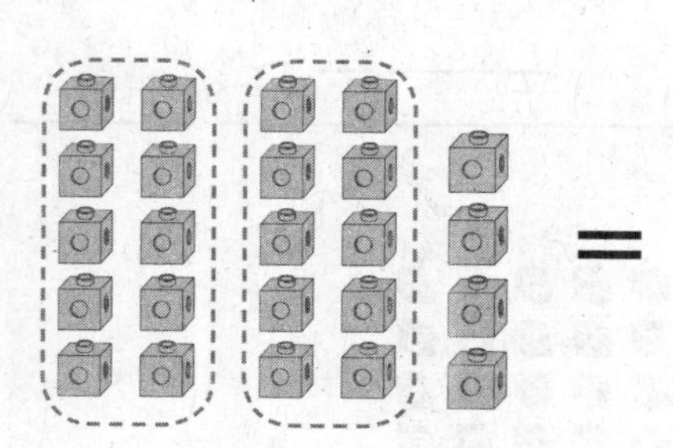

=

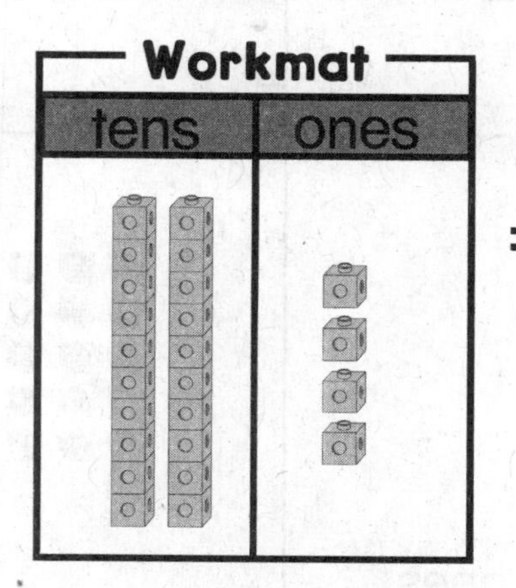

=

tens	ones
2	4

_2_ tens _4_ ones is the same as _24_. There are _24_ in all.

Use Workmat 3 and [cube].
Write how many tens and ones. Write the number.

1.

___ tens ___ ones

___

2.

___ tens ___ ones

___

Name ____________________ **Lesson 3.2**

Hands On: Tens and Ones

Use Workmat 3 and ▪. Write how many tens and ones.
Then write the number.

1.

2 tens 9 ones

29

2.

____ tens ____ ones

____

3.

____ tens ____ ones

____

4.

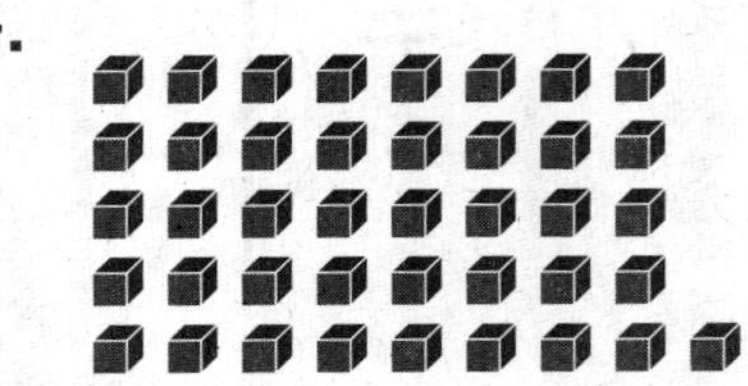

____ tens ____ ones

____

Problem Solving

5. Ricky has 5 bags of small marbles and 3 big marbles. Each bag has 10 marbles. How many marbles does he have?

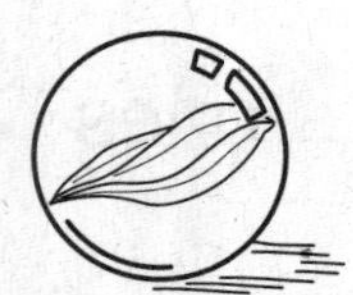

marble

____ marbles

6. Tad has 27 cookies. How many boxes of 10 cookies can he make?

____ boxes

How many cookies will not be in a box?

____ cookies

cookie

Name ____________________

Lesson 3.3

Understand Place Value

0, 1, 2, 3, 4, 5, 6, 7, 8, and 9 are **digits**.
When you write 52, you use two digits.
The first digit is 5. The second digit is 2.
Digits can mean different things in different numbers. You can tell what a different digit means by looking at its place in a number.

52

The digit 5 is in the tens place.
It tells you that 52 has 5 tens.
The digit 5 in 52 means 50.

The digit 2 is in the ones place.
It tells you that 52 has 2 ones.
The digit 2 in 52 means 2.

Circle what the gray digit means.

1. 27

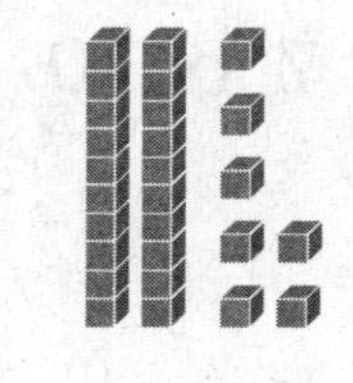

20 or 2

2. 18

I or 10

3. 56

60 or 6

4. 30

30 or 3

5. 75

5 or 50

6. 41

40 or 4

Name____

Understand Place Value

Circle the value of the gray digit.

1. 36 3 or 30	2. 52 50 or 5	3. 60 6 or 60
4. 15 1 or 10	5. 86 80 or 8	6. 53 3 or 30
7. 72 2 or 20	8. 75 70 or 7	9. 24 2 or 20
10. 18 8 or 80	11. 19 10 or 1	12. 93 9 or 90
13. 85 5 or 50	14. 38 30 or 3	15. 67 7 or 70

Problem Solving

16. I have 4 ones and 3 tens. What number am I?

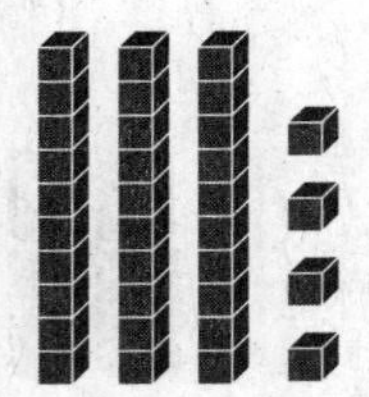

______

17. I have 8 ones and 1 ten. What number am I?

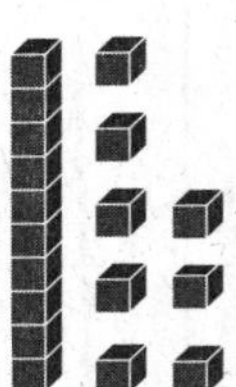

______

Spiral Review

Write **is greater than**, **is less than**, or **is equal to**.
Then write >, <, or =.

1. 47 ______________________ 63.

   47 ◯ 63

2. 28 ______________________ 23.

   28 ◯ 23

Find the pattern. Complete the table to solve.

3. How many wheels are on 5 cars?

number of cars	1	2	3	4	5
number of wheels	4	8	12		

There are _____ wheels on each car.

There are ________ wheels on 5 cars.

Use [balance] and the real object. Choose a unit to measure.
Draw it. Estimate. Then measure.

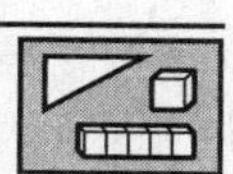

4.

Object	Unit	Estimate	Measurement
		about ________	about ________

Draw a picture of the problem.
Write the number sentence. Solve.

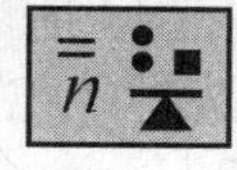

5. Neil bakes 12 muffins. He gives 4 of them away. How many muffins does Neil have left?

_____ ◯ _____ ◯ _____

Name____________________ **Lesson 3.6**

Different Ways to Show Numbers

Here are some different ways to show 28.

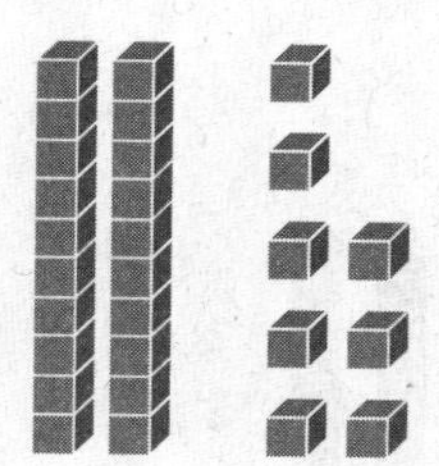

There are 2 tens.
There are 8 ones.
There are 28 in all.

You can show 28 with 2 tens and 8 ones.

There is 1 ten.
There are 18 ones.
There are 28 in all.

You can show 28 with 1 ten and 18 ones.

There are 0 tens.
There are 28 ones.
There are 28 in all.

You can show 28 with 0 tens and 28 ones.

Use [tens block] [ones cube] if you need to. Write how many tens and ones.

1. 32

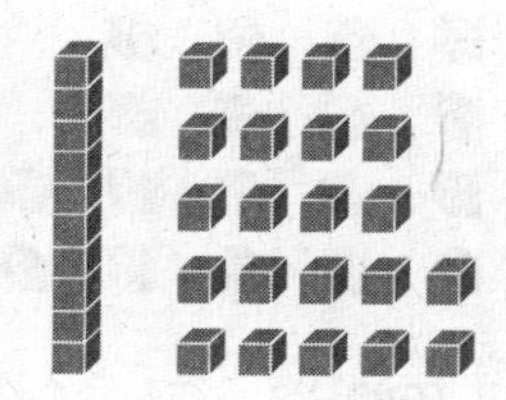

1 ten 22 ones ____ tens ____ ones ____ tens ____ ones

2. 40

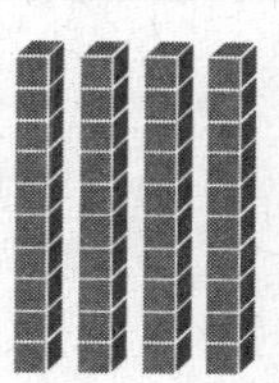

____ tens ____ ones ____ tens ____ ones ____ tens ____ ones

Name ____________________

Lesson 3.6

Different Ways to Show Numbers

Use tens and ones blocks. Write how many tens and ones.

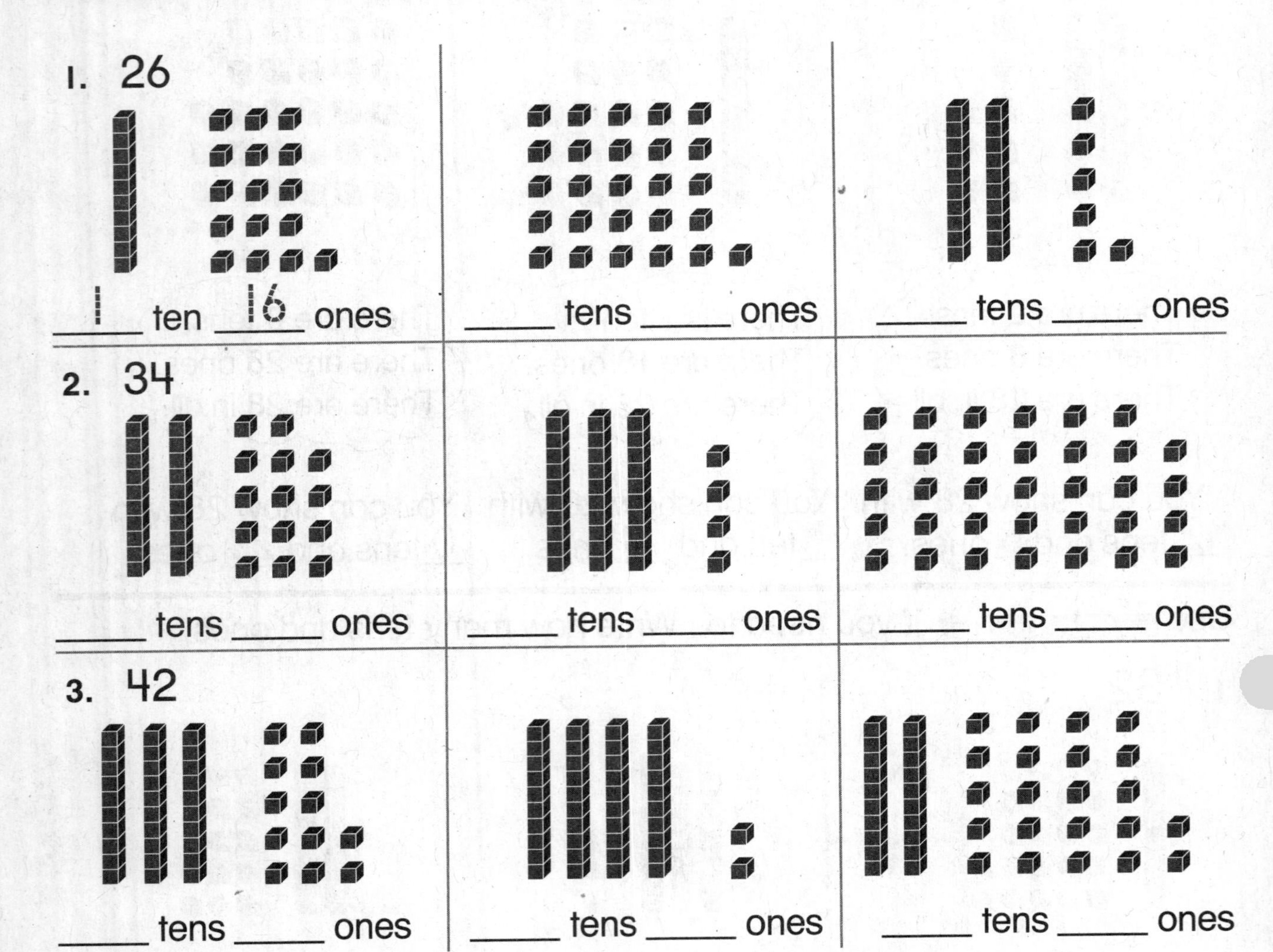

Problem Solving

4. Use tens and ones blocks. Draw two ways to show 37.
Write how many tens and ones for each model.

_______ tens _______ ones _______ tens _______ ones

Name ______________________________ **Week 6**

Spiral Review

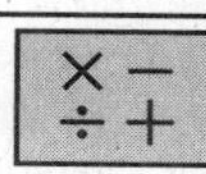

Write two ways to describe the meaning of the number.

1.

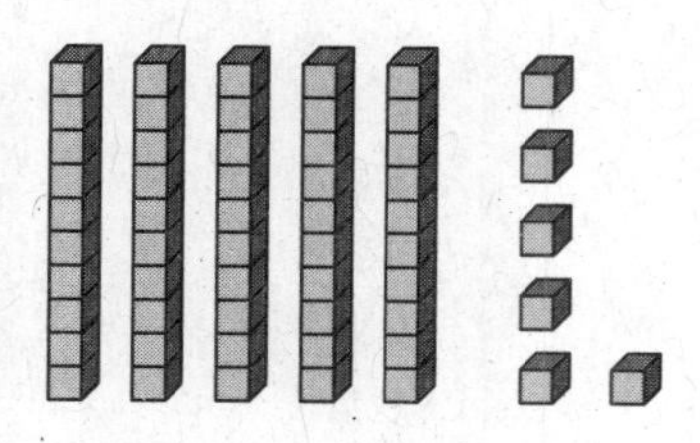

56 = ______ tens ______ ones

56 = ______ + ______

Skip-count. Show the pattern on the hundred chart.

2. Count by twos.
Color the squares red.

3. Count by fives.
Circle the numbers.

1	2	3	4	5	6	7	8	9	10
11	12	13	14	15	16	17	18	19	20
21	22	23	24	25	26	27	28	29	30
31	32	33	34	35	36	37	38	39	40
41	42	43	44	45	46	47	48	49	50
51	52	53	54	55	56	57	58	59	60
61	62	63	64	65	66	67	68	69	70
71	72	73	74	75	76	77	78	79	80

Circle each cylinder.

4.

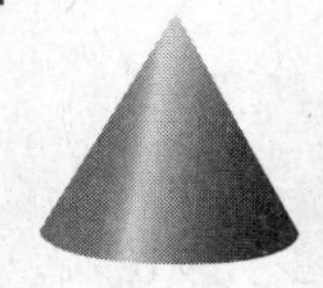      

Use Workmat 11 and [tens and ones blocks]. If you can make a ten, regroup. Write how many tens and ones. Write the sum.

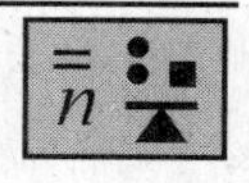

5. Add 34 and 9.

Workmat

Tens	Ones

______ tens ______ ones

______

Name____________________________________ **Lesson 4.2**

Order Numbers

You can put three numbers in order.

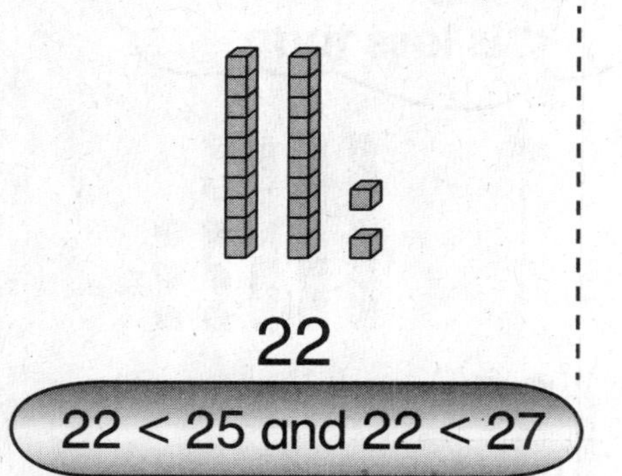

22

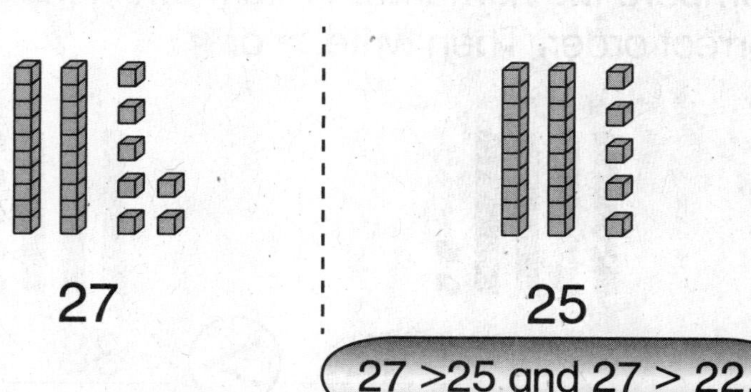

27

25

22 < 25 and 22 < 27

27 >25 and 27 > 22.

The **least** number is 22. It is less than both other numbers.

The **greatest** number is 27. It is greater than both of the other numbers.

You can order the numbers in two ways.

You can order them from least to greatest.

22 (<) 25  27

least **greatest**

You can order them from greatest to least.

27 (>) 25 (>) 22

greatest **least**

Compare the numbers. Write the correct order. Then write > or <.

1. 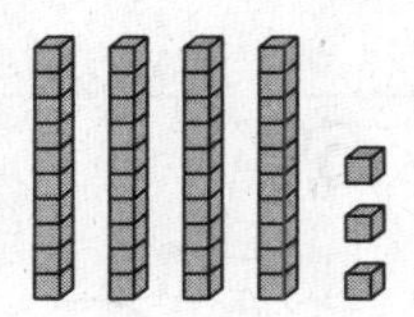 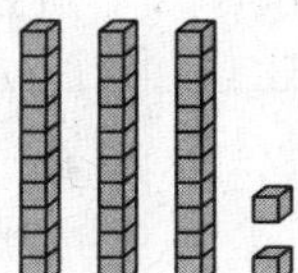

____ ◯ ____ ◯ ____

least **greatest**

2. 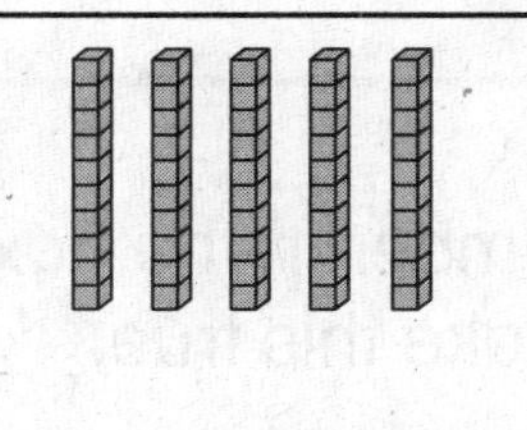 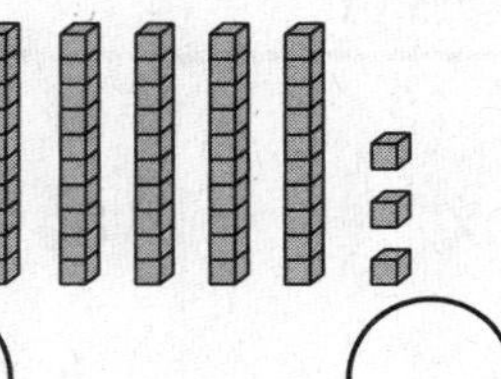

____ ◯ ____ ◯ ____

greatest **least**

Name ____________________

Order Numbers

Compare the numbers. Write them in the correct order. Then write > or <.

REMEMBER:
> is greater than
< is less than

1.

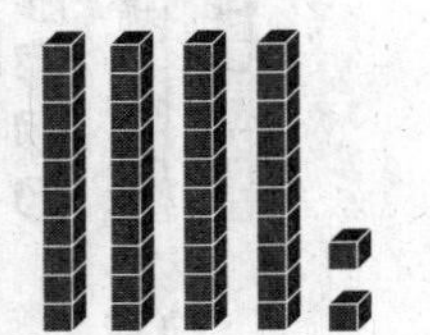

42 (>) 38 (>) 35

greatest — least

2.

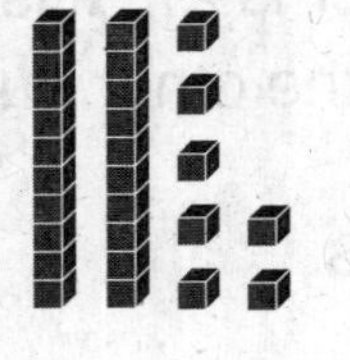

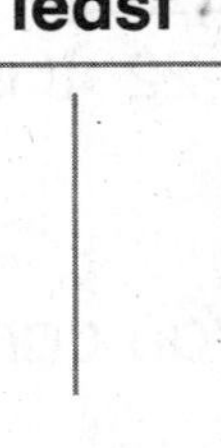

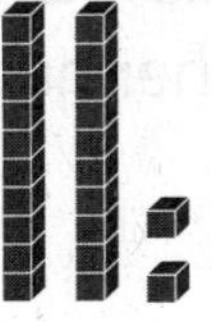

____ ◯ ____ ◯ ____

least — greatest

Write the numbers in the correct order. Then write > or <.

3. 45 33 61

____ ◯ ____ ◯ ____

least — greatest

4. 56 55 64

____ ◯ ____ ◯ ____

greatest — least

5. 18 37 46

____ ◯ ____ ◯ ____

greatest — least

6. 60 59 63

____ ◯ ____ ◯ ____

least — greatest

Problem Solving

7. Write a number in the box that will make this true.

41 < ☐ < 44

8. Write a number in the box that will make this true.

61 > ☐ > 56

Name ______________________________ **Week 7**

Spiral Review

Add.
Regroup if you need to.

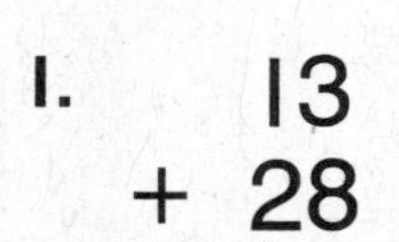

1. $13 + 28$

2. $25 + 31$

3. $79 + 11$

4. $66 + 15$

Use cubes to show the number as tens and ones. Draw what you built. Write **even** or **odd**.

5. 36

____

Cross off figures that do not match the clues.
Circle the figure that is left.

6. This figure has straight sides.
It has more than 3 sides.
It has 4 corners.
Which is the figure?

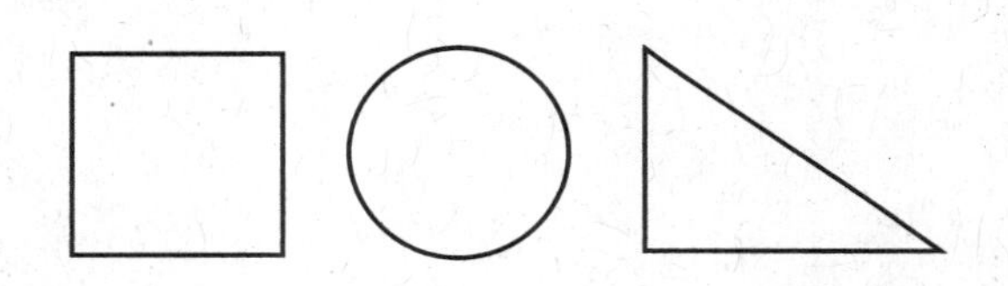

7. I have no curves.
I have more than 5 corners.
My sides are not all the same.
Which figure am I?

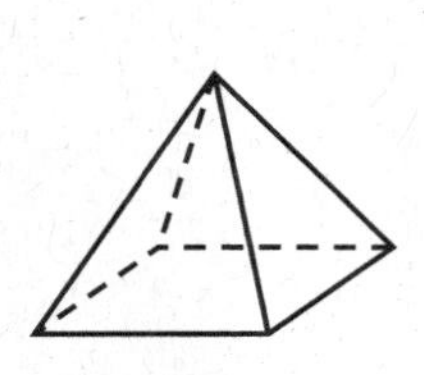
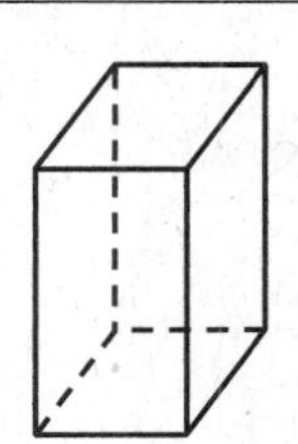
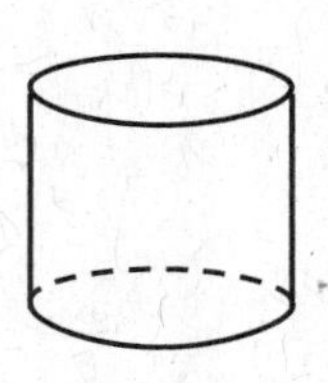

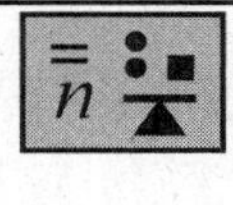

Find the sums.

8. $4 + 7 =$ ____
$7 + 4 =$ ____

9. ____ $= 0 + 4$
____ $= 0 + 7$

10. $8 + 5 =$ ____
$8 + 6 =$ ____

Name____

Algebra: Skip-Count on a Number Line

18, 22, 26, _____, _____

What numbers come next?

Use a number line. Draw the jumps.

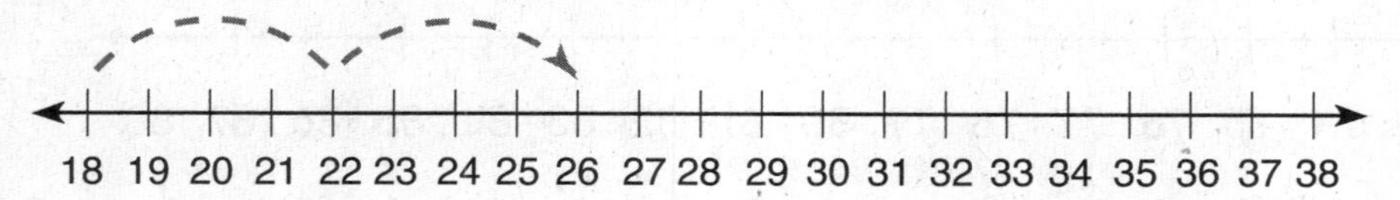

Each jump adds 4. You are skip-counting by fours.

Extend the pattern.

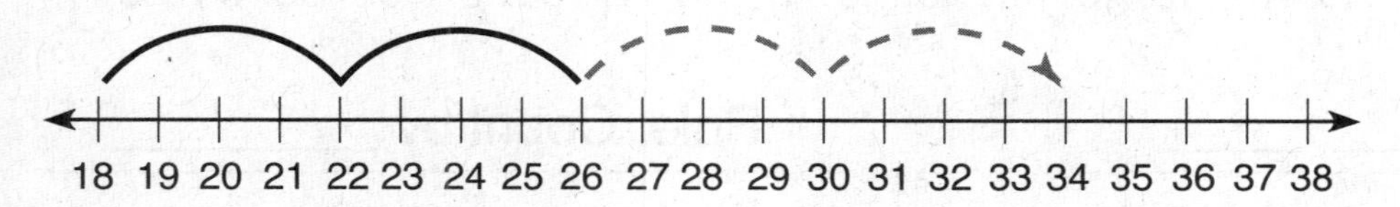

18, 22, 26, 30, 34

The rule is: Count by fours.

Skip-count. Write the missing numbers.
Write a rule for the pattern.

1.

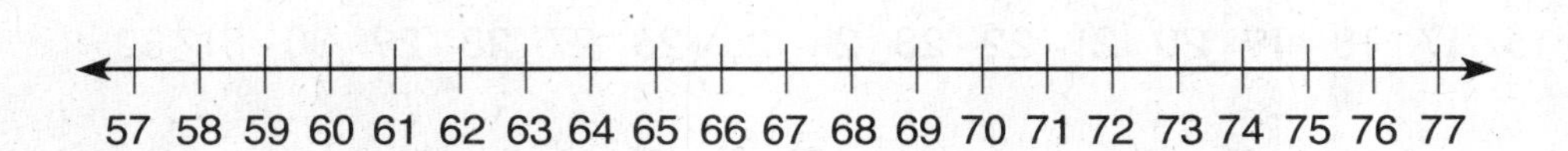

57, 59, 61, _____, _____ Rule: Count by _____.

2.

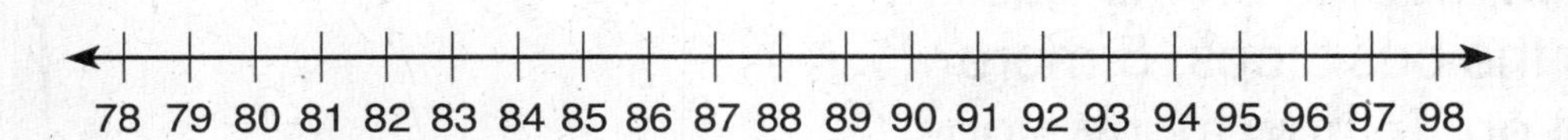

78, 83, 88, _____, _____ Rule: Count by _____.

SDAP2.2 – Solve problems involving simple number patterns. Students demonstrate an understanding of patterns and how patterns grow and describe them in general ways.

Name ____________________

Algebra: Skip-Count on a Number Line

Skip-count. Write the missing numbers.
Write a rule for the pattern.

1.

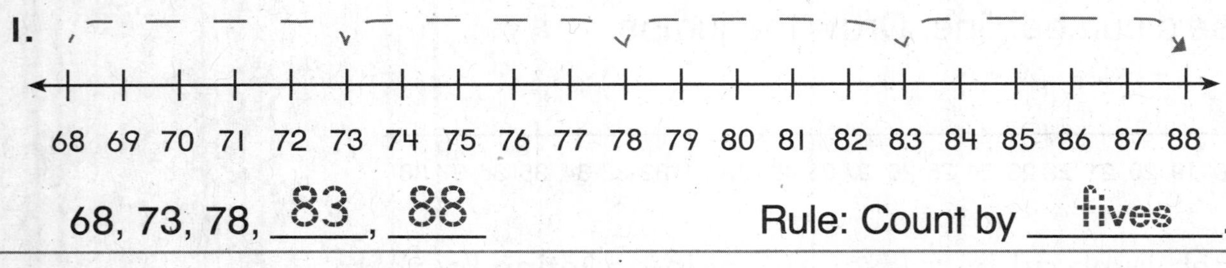

68, 73, 78, 83, 88 — Rule: Count by fives.

2.

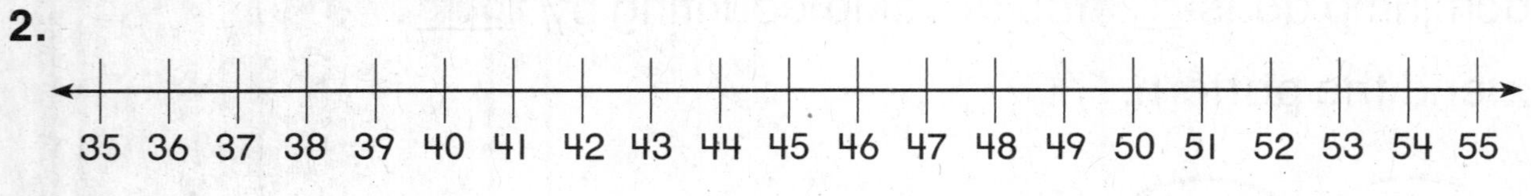

35, 37, 39, _____, _____ — Rule: Count by __________.

3.

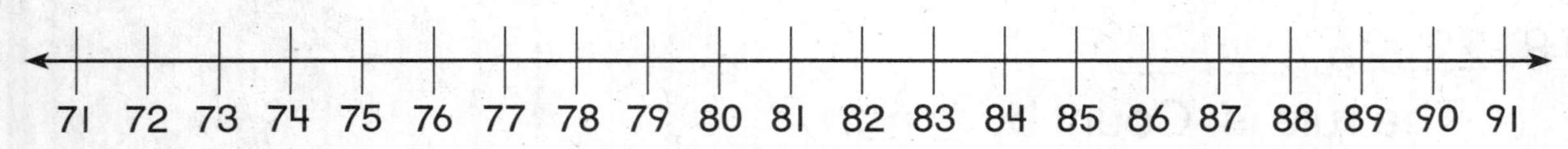

71, 75, 79, _____, _____ — Rule: Count by __________.

4.

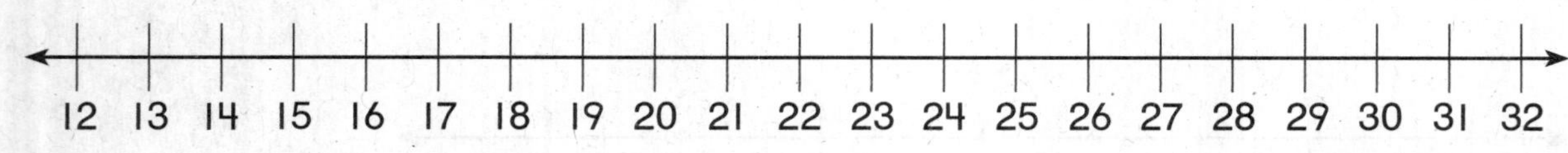

12, 15, 18, _____, _____ — Rule: Count by __________.

Problem Solving

37 38 39 40 41 42 43 44 45 46 47 48 49 50 51 52 53 54 55 56 57

5. There are 37 people on the bus. Each time the bus stops, 5 more people get on. In how many stops will there be 57 people on the bus?

_____ stops

Name____________________

Lesson 4.6

Problem Solving Strategy: Find a Pattern

How many wheels are on 5 scooters?

Read to Understand

•• What do you want to find out?
How many wheels there are on 5 scooters.

Plan

•• How will you solve this problem? find a pattern

Solve

•• Complete the table. The pattern is count by twos.

number of scooters	1	2	3	4	5
number of wheels	2	4	6	8	10

There are 2 wheels on each scooter.

There are 10 wheels on 5 scooters.

Check

•• Does your answer make sense? Yes. If I add 2 five times, I get 10. $2 + 2 + 2 + 2 + 2 = 10$

Find the pattern. Complete the table to solve.

•• How many pedals are on 6 bicycles?

number of bicycles	1	2	3	4	5	6
number of pedals	2	4	6			

There are ___ pedals on each bicycle.

There are ____ pedals on 6 bicycles.

•• How many sails are on 5 boats?

number of boats	1	2	3	4	5
number of sails	5	10	15		

There are ___ sails on each boat.

There are ____ sails on 5 boats.

SDAP2.2 — Solve problems involving simple number patterns. **SDAP2.0** — Students demonstrate an understanding of patterns and how patterns grow and describe them in general ways. **SDAP2.1** — Recognize, describe, and extend patterns and determine a next term in linear patterns (e.g., 4, 8, 12... ; The number of ears on one horse, two horses, three horses, four horses). **MR2.0** — Students solve problems and justify their reasoning. **MR2.1** — Defend the reasoning used and justify the procedure selected.

Name____

Lesson 4.6

Problem Solving Strategy: Find a Pattern

Find the pattern. Complete the table to solve.

1. How many wings are on 6 airplanes?

number of airplanes	1	2	3	4	5	6
number of wings	2	4	6			

There are _____ wings on each airplane.

There are _____ wings on 6 airplanes.

2. How many wheels are on 5 wagons?

number of wagons	1	2	3	4	5
number of wheels	4	8	12		

There are _____ wheels on each wagon.

There are _____ wheels on 5 wagons.

Mixed Strategy Practice

Choose a strategy to solve.

Choose a Strategy

- Draw a Picture
- Write a Number Sentence
- Find a Pattern

3. Matt has 6 black socks. He has 4 white socks. How many socks does he have in all?

sock 

_____ socks

4. Tasha has 14 grapes. Seven are purple. The rest are green. How many grapes are green?

grape 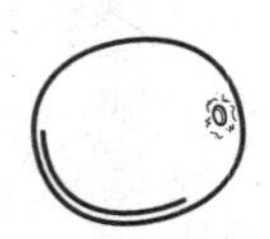

_____ grapes

Name ________________________ Lesson 4.7

Algebra: Number Patterns

You can start patterns with different numbers.

Skip-count by fives.

Start at 3. Shade 3.

Count by 5. Shade 8.

Count by 5. Shade 13.

1	2	3	4	5	6	7	8	9	10
11	12	13	14	15	16	17	18	19	20
21	22	23	24	25	26	27	28	29	30
31	32	33	34	35	36	37	38	39	40
41	42	43	44	45	46	47	48	49	50
51	52	53	54	55	56	57	58	59	60
61	62	63	64	65	66	67	68	69	70
71	72	73	74	75	76	77	78	79	80
81	82	83	84	85	86	87	88	89	90
91	92	93	94	95	96	97	98	99	100

Extend the pattern:

3, 8, 13, 18, 23, 28, 33, 38

Use the hundred chart. Extend the pattern.

1. Skip-count by twos.

7, 9, 11, 13, 15, 17, 19, 21

2. Skip-count by tens.

13, 23, 33, ____, ____, ____, ____, ____

3. Skip-count by fives.

16, 21, 26, ____, ____, ____, ____, ____

4. Skip-count by twos.

41, 43, 45, ____, ____, ____, ____, ____

SDAP2.0 — Students demonstrate an understanding of patterns and how patterns grow and describe them in general ways.

Name ______________________

Lesson 4.7

Algebra: Number Patterns

Use the hundred chart.
Extend the pattern.

1	2	3	4	5	6	7	8	9	10
11	12	13	14	15	16	17	18	19	20
21	22	23	24	25	26	27	28	29	30
31	32	33	34	35	36	37	38	39	40
41	42	43	44	45	46	47	48	49	50
51	52	53	54	55	56	57	58	59	60
61	62	63	64	65	66	67	68	69	70
71	72	73	74	75	76	77	78	79	80
81	82	83	84	85	86	87	88	89	90
91	92	93	94	95	96	97	98	99	100

1. Skip-count by fives.

4, 9, 14, 19, 24, 29, 34, 39

2. Skip-count by tens.

8, 18, 28, ____, ____, ____, ____, ____

3. Skip-count by twos.

9, 11, 13, ____, ____, ____, ____, ____

4. Skip-count by fives.

31, 36, 41, ____, ____, ____, ____, ____

5. Skip-count by tens.

15, 25, 35, ____, ____, ____, ____, ____

Problem Solving

6. Gene is skip-counting by tens. He starts the pattern with 26. Will he say 86 in his pattern? Circle **yes** or **no.**

yes no

7. Write the first five numbers in Gene's pattern.

____, ____, ____, ____, ____

Name______________________________

Lesson 5.1

Mental Math: Add on Multiples of Ten

What is 35 + 30?
Use a hundred chart to count on by tens.

Start on 35.

Move down three rows.
Each row adds a ten.

Count on 3 tens:

35, 45, 55, 65

So, 35 + 30 = 65

1	2	3	4	5	6	7	8	9	10
11	12	13	14	15	16	17	18	19	20
21	22	23	24	25	26	27	28	29	30
31	32	33	34	**35**	36	37	38	39	40
41	42	43	44	**45**	46	47	48	49	50
51	52	53	54	**55**	56	57	58	59	60
61	62	63	64	**65**	66	67	68	69	70
71	72	73	74	75	76	77	78	79	80
81	82	83	84	85	86	87	88	89	90
91	92	93	94	95	96	97	98	99	100

REMEMBER: **30** is the same as **3** tens.

Count on to add.

1. 23 + 30 = 53
2. 19 + 50 = ____
3. 62 + 20 = ____
4. 41 + 30 = ____
5. 39 + 40 = ____
6. 52 + 40 = ____
7. 15 + 20 = ____
8. 34 + 50 = ____

NS2.3 Use mental arithmetic to find the sum or difference of two two-digit numbers. **NS2.2** Find the sum or difference of two whole numbers up to three digits long. **NS2.0** Students estimate, calculate, and solve problems involving addition and subtraction of two- and three-digit numbers.

Name ______________________________

Mental Math: Add on Multiples of Ten

Count on to add.

1. $36 + 40 =$ 76	2. $54 + 20 =$ ____
3. $27 + 30 =$ ____	4. $78 + 20 =$ ____
5. $14 + 50 =$ ____	6. $35 + 30 =$ ____
7. $17 + 10 =$ ____	8. $43 + 40 =$ ____
9. $21 + 40 =$ ____	10. $46 + 50 =$ ____

Problem Solving

11. Ben has 22 owl stickers. His friend gives him some more. Now he has 62 owl stickers. Count on by tens. How many owl stickers did Ben's friend give him?

owl sticker

____ owl stickers

12. Jada has 16 colored markers. She buys some more. Now she has 46 markers. Count on by tens. How many markers did Jada buy?

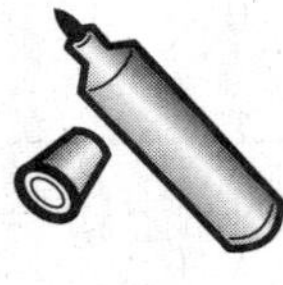

marker

____ markers

Spiral Review

Break apart the addends. Find the sum.

1. 28 + 64 = ?

____ + ____ + ____ + ____

Add the tens. ____ + ____ = ____

Add the ones. ____ + ____ = ____

How many in all? ____ + ____ = ____

So, 28 + 64 = ____.

Skip-count. Write the missing numbers.
Write a rule for the pattern.

2. 19 20 21 22 23 24 25 26 27 28 29 30 31 32 33 34 35 36 37 38 39

19, 23, 27, ____, ____ Rule: Count by ______.

Cross out the paintings that do not match the clues.
Circle the answer.

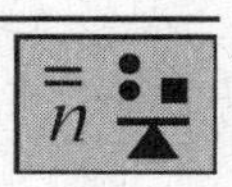

3. Sam paints a circle above a square. He paints a triangle to the left of the circle. Which painting is Sam's?

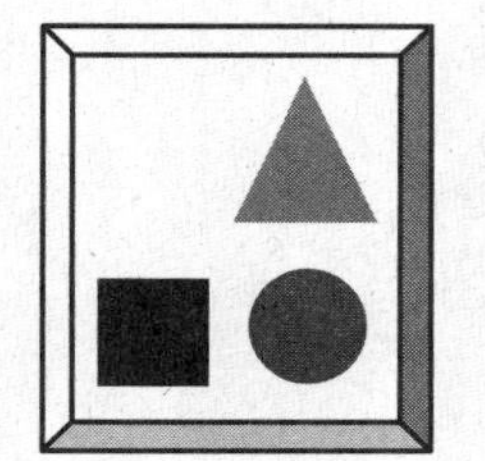

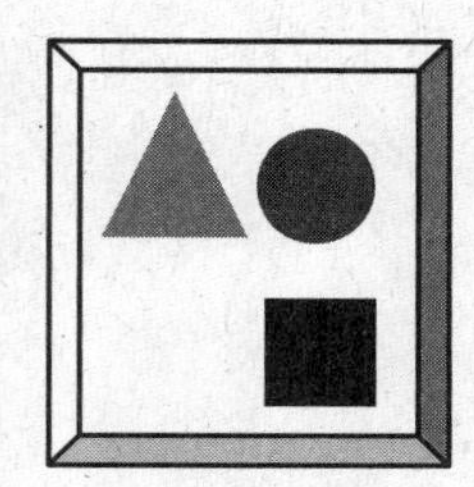

Use the table to solve.

Points Scored in the Game	
Player	**Number of Points**
Kim	16
Terence	15
Joe	27

4. How many points did Kim and Terence score in all?

____ points

Name ______________________

Lesson 5.2

Regrouping for Addition

Add 37 and 4.

Show 37 and 4.

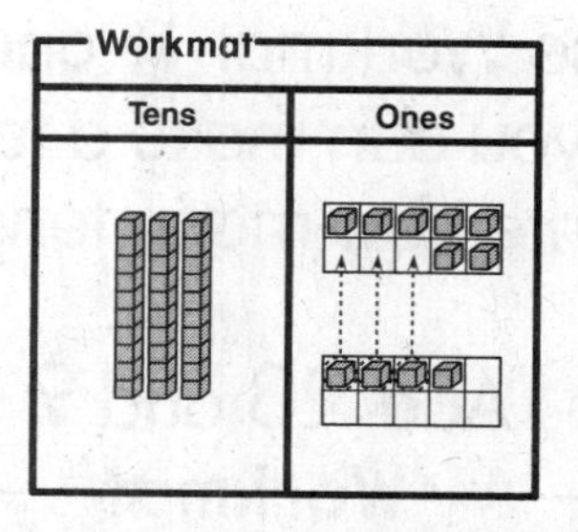

Count the ones.
How many ones are there in all? __11__
Can you make a ten? __yes__

Trade 10 ones
for 1 ten.
This is called
regrouping.

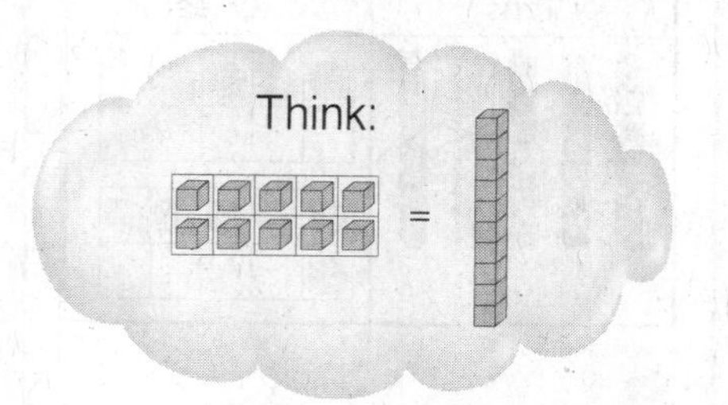

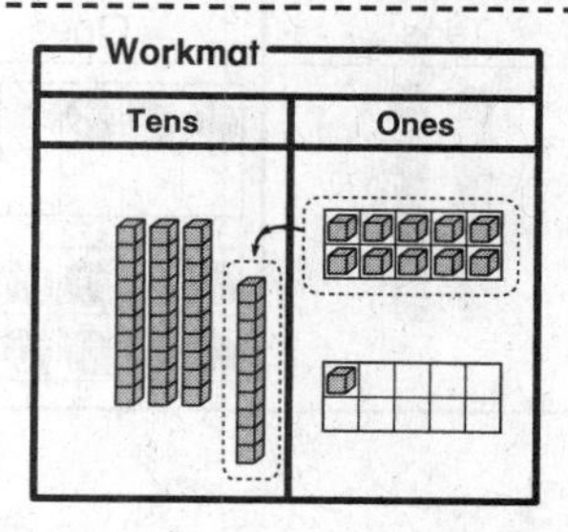

Count the tens. How many tens are there in all? __4__ tens
Count the ones. How many ones are there in all? __1__ one

__4__ tens __1__ one is the same as __41__.

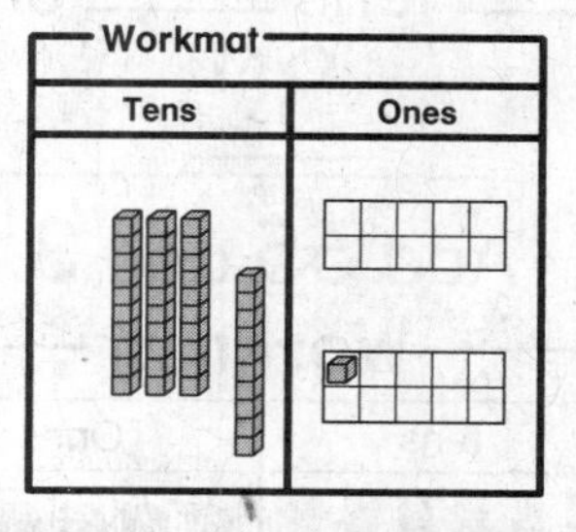

Use Workmat 11 and .
If you can make a ten, regroup.
Write how many tens and ones. Write the sum.

1. Add 58 and 5.

Workmat: Tens | Ones

__6__ tens __3__ ones

__63__

2. Add 29 and 6.

Workmat: Tens | Ones

_____ tens _____ ones

_____

3. Add 47 and 9.

Workmat: Tens | Ones

_____ tens _____ ones

_____

NS 2.2 Find the sum or difference of two whole numbers up to three digits long. **AF 1.0** Students model, represent, and interpret number relationships to create and solve problems involving addition and subtraction.

Name ______________________________

Lesson 5.2

Hands On: Regrouping for Addition

Use Workmat 11 and [tens block] [ones block].
If you can make a ten, regroup.
Write how many tens and ones. Write the sum.

1. Add 23 and 9.

Workmat

Tens	Ones

3 tens 2 ones

32

2. Add 54 and 6.

Workmat

Tens	Ones

_____ tens _____ ones

_____

3. Add 37 and 7.

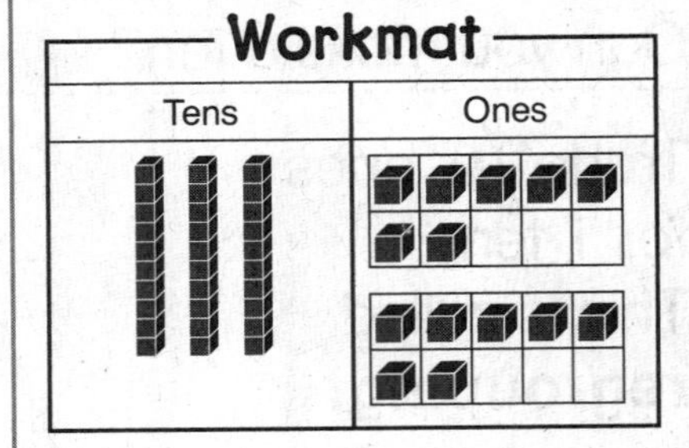

_____ tens _____ ones

_____

4. Add 65 and 3.

Workmat

Tens	Ones

_____ tens _____ ones

_____

5. Add 18 and 6.

Workmat

Tens	Ones

_____ tens _____ ones

_____

6. Add 42 and 5.

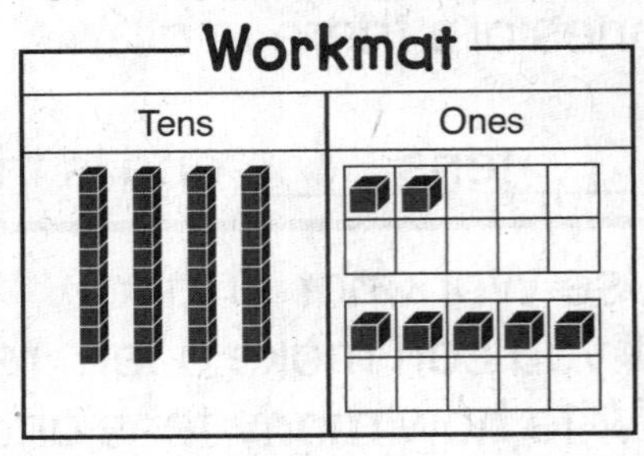

_____ tens _____ ones

_____

Problem Solving

Use Workmat 11 and [tens block] [ones block] to solve.

7. Shelby sees 34 owls at the zoo. Then she sees 9 more owls on a nature walk. How many owls does Shelby see in all?

_____ owls

owl

Name ____________________

Model 2-Digit Addition

Add 18 and 25.
Show 18 and 25 with [tens] [ones].
Count the ones.
How many ones are there in all? 13
Can you make a ten? yes

Trade 10 ones for 1 ten.
This is called regrouping.

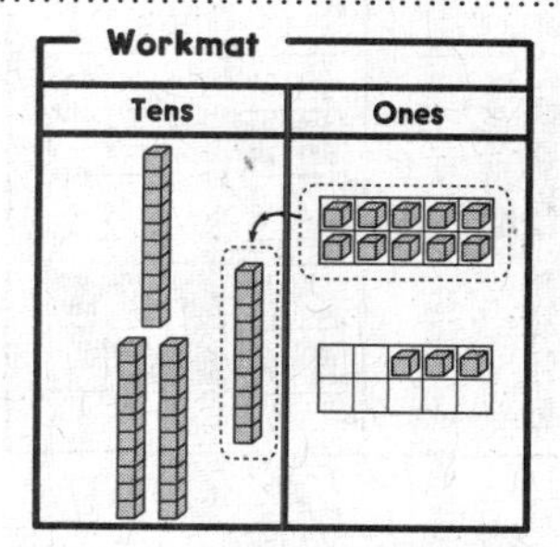

Count the tens. How many tens are there in all? 4 tens

Count the ones. How many ones are there in all? 3 ones

4 tens 3 ones is the same as 43.

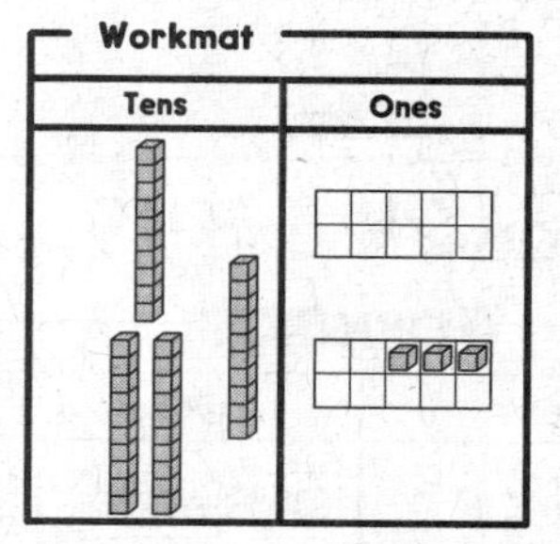

Workmat 11 and [tens] [ones].
If you can make a ten, regroup.
Write how many tens and ones. Write the sum.

1. Add 46 and 19.

Workmat — Tens | Ones

__ tens __ ones

____

2. Add 45 and 27.

Workmat — Tens | Ones

__ tens __ ones

____

3. Add 58 and 38.

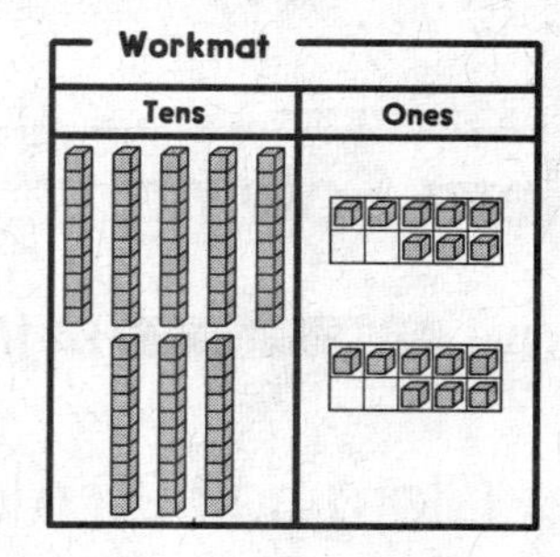

__ tens __ ones

____

Name ______________________________

Lesson 5.3

Hands On: Model 2-Digit Addition

Use Workmat 11 and [ten-block] [one-block].
If you can make a ten, regroup.
Write how many tens and ones. Write the sum.

1. Add 33 and 19.

Workmat

Tens	Ones

5 tens 2 ones

52

2. Add 46 and 37.

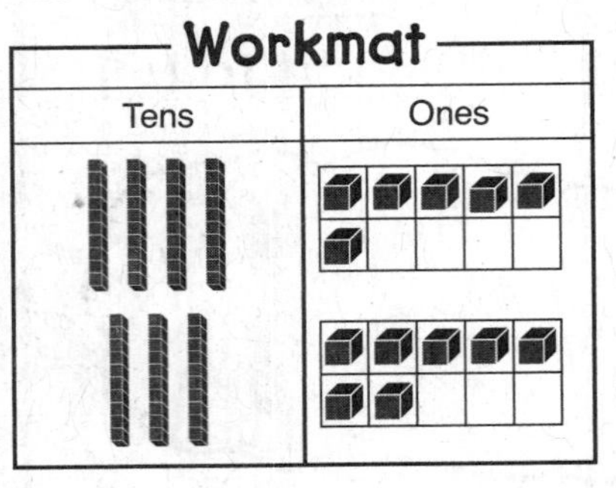

____ tens ____ ones

____

3. Add 51 and 28.

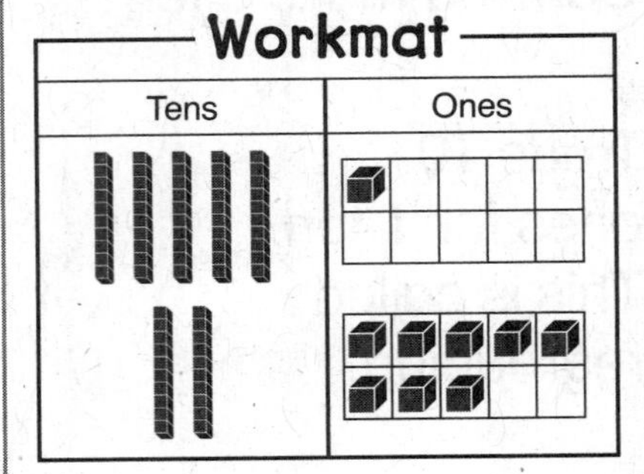

____ tens ____ ones

____

4. Add 24 and 49.

Workmat

Tens	Ones

____ tens ____ ones

____

5. Add 12 and 77.

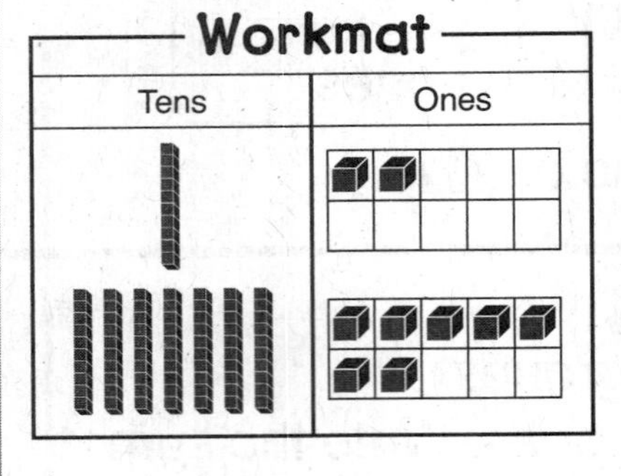

____ tens ____ ones

____

6. Add 68 and 29.

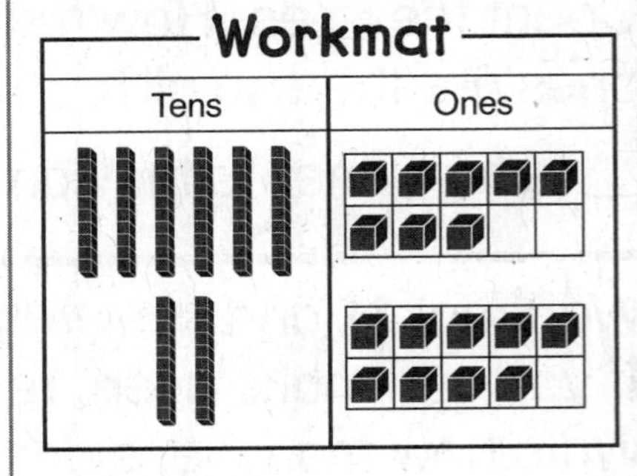

____ tens ____ ones

____

Problem Solving

- Draw [ten-block] [one-block] to solve.

7. Clara has 25 marbles. Keith has 18 marbles. How may marbles do Clara and Keith have in all?

marble

____ marbles

Name ______________________

Spiral Review

Subtract. Add to check.

1.

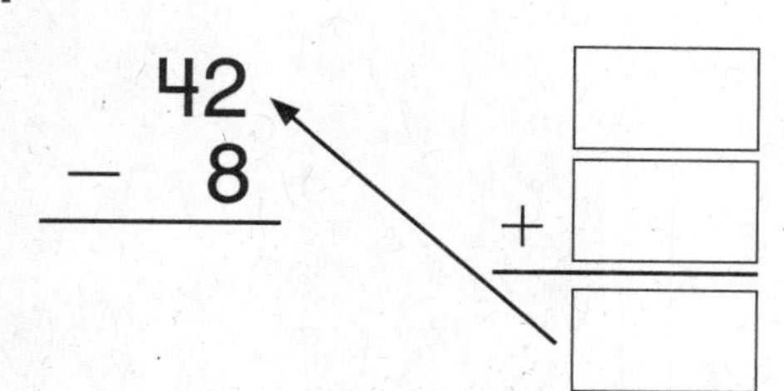

Use cubes to show the number as tens and ones. Draw what you built. Write **even** or **odd**.

2. 27

______

Circle every shape that makes the statement true.

3. The ______ is to the **left of** the ▲.

4. The ______ is **above** the ☐.

Use the table to solve.

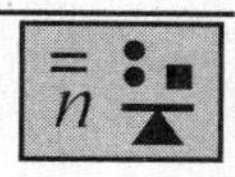

Crayons in a Box	
Color	**Number of Crayons**
red	32
blue	47
green	28
yellow	19

5. Alex takes 15 red crayons out of the box. How many red crayons are still in the box?

______ red crayons

Name____ **Lesson 6.3**

Rewrite 2-Digit Addition

Add. 43 + 19 = ?

STEP 1

What is the tens digit in 43? __4__
Write 4 in the tens column.
Write the ones digit in the ones column.

Tens	Ones
☐	
4	3
+	

STEP 2

What is the tens digit in 19? __1__
Write 1 in the tens column.
Write the ones digit in the ones column.

Tens	Ones
☐	
4	3
+ 1	9

STEP 3

Add the ones.
Regroup if you need to.
Add the tens.

Tens	Ones
1	
4	3
+ 1	9
6	2

Rewrite the numbers. Then add.

1. 26 + 9

Tens	Ones
1	
2	6
+	9
3	5

2. 32 + 38

3. 16 + 57

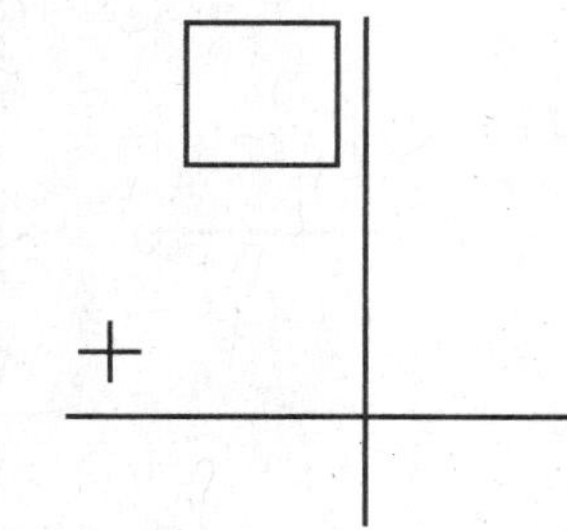

4. 24 + 26

5. 17 + 62

6. 34 + 8

7. 22 + 49

8. 54 + 29

Name___________________________ **Lesson 6.3**

Rewrite 2-Digit Addition

Rewrite the numbers. Then add.

1. 35 + 57

	Tens	Ones
	1	
	3	5
+	5	7
	9	2

2. 22 + 19

	Tens	Ones
+		

3. 43 + 6

	Tens	Ones
+		

4. 16 + 56

	Tens	Ones
+		

5. 21 + 24

6. 52 + 38

7. 34 + 48

8. 29 + 7

9. 33 + 27

10. 19 + 26

11. 44 + 15

12. 12 + 29

Problem Solving

13. Erik hit 36 tennis balls in the morning and 47 tennis balls in the afternoon. His father hit 82 tennis balls in all. Who hit more tennis balls?

______

Spiral Review

Look at the difference in the box.
Write > or < to make each sentence true.

1. $60 - 20 = 40$

$60 - 17$ ◯ 40

$60 - 22$ ◯ 40

2. $50 - 30 = 20$

$50 - 34$ ◯ 20

$50 - 31$ ◯ 20

3. Use the picture to complete the table.
Then shade bars in the graph to show the data

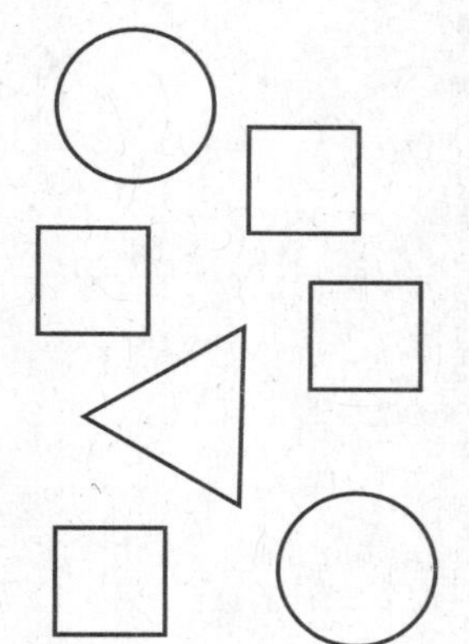

Figures	
Figure	**Number**
circle	
square	
triangle	

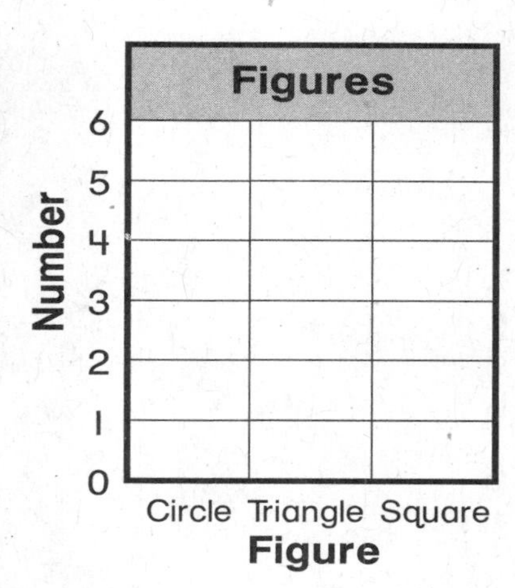

Use solids.

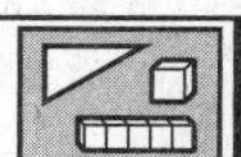

4. Circle each solid with both curves and flat surfaces.

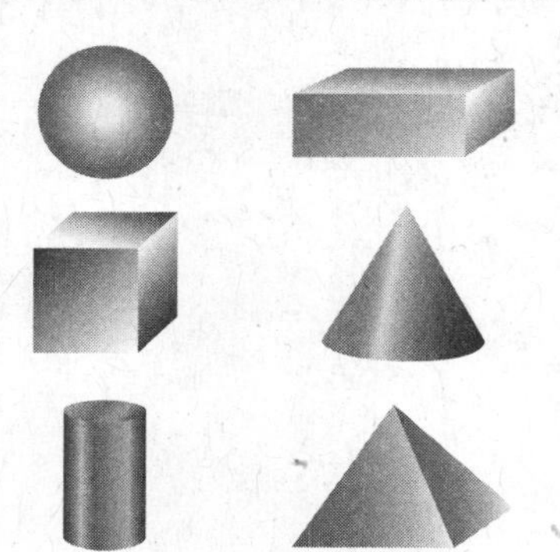

Follow the rule.
Complete the table.

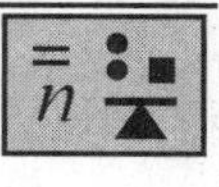

5. Rule: Add 4

In	Out
3	
5	
7	
9	

Name____________________________________

Lesson 6.4

Estimate Sums

Sums that you know can help you estimate other sums.

Look at the sum in the box. **40 + 50 = 90**

Write > or < to make the sentence true. 39 + 52 ◯ 90

Compare the addends. 39 < 40 and 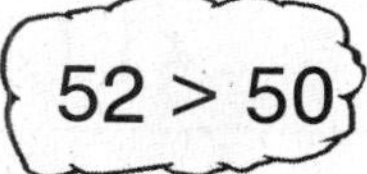

One addend is 1 less and the other addend is 2 more. 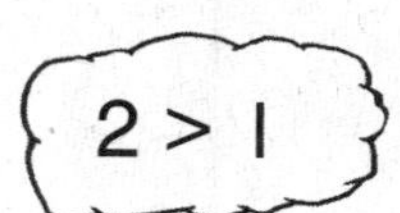

So, 39 + 52 (>) 90

Look at the sum in the box.
Write > or < to make each sentence true.

1. **20 + 60 = 80**

22 + 60 (>) 80

19 + 59 (<) 80

2. **40 + 40 = 80**

37 + 38 ◯ 80

42 + 39 ◯ 80

3. **70 + 20 = 90**

70 + 19 ◯ 90

68 + 23 ◯ 90

4. **60 + 30 = 90**

62 + 29 ◯ 90

58 + 33 ◯ 90

5. **10 + 30 = 40**

11 + 31 ◯ 40

9 + 32 ◯ 40

6. **20 + 20 = 40**

17 + 22 ◯ 40

23 + 19 ◯ 40

NS2.0 Students estimate, calculate, and solve problems involving addition and subtraction of two-and three-digit numbers. **NS6.0** Students use estimation strategies in computation and problem solving that involve numbers that use the ones, tens, hundreds, and thousands places. **MR3.0** Students note connections between one problem and another.

Name ____________________

Estimate Sums

Look at the sum in the box.
Write > or < to make each sentence true.

1. $10 + 50 = 60$

$8 + 48$ (<) 60

$11 + 50$ (>) 60

2. $20 + 40 = 60$

$21 + 38$ ( ) 60

$20 + 42$ ( ) 60

3. $70 + 20 = 90$

$73 + 20$ ( ) 90

$68 + 18$ ( ) 90

4. $40 + 50 = 90$

$40 + 53$ ( ) 90

$38 + 48$ ( ) 90

5. $50 + 20 = 70$

$47 + 19$ ( ) 70

$52 + 20$ ( ) 70

6. $30 + 40 = 70$

$32 + 42$ ( ) 70

$30 + 39$ ( ) 70

Problem Solving

7. Choose from the numbers on the flag to write true sentences. Use the sum in the box to help.

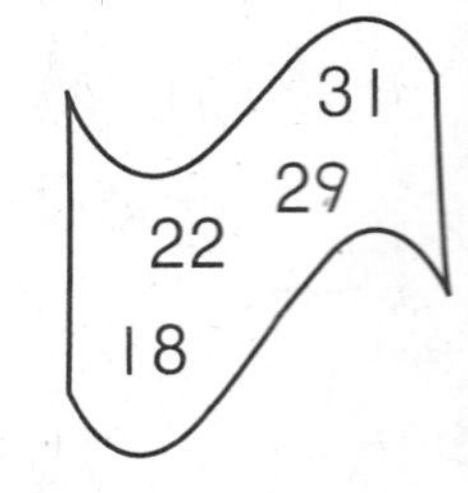

____ + ____ > 50 ____ + ____ < 50

____ + ____ > 50 ____ + ____ < 50

____ + ____ > 50 ____ + ____ < 50

Name ____________________

Lesson 6.5

More 2-Digit Addition

Eliza sold 47 soccer balls in one week.
She sold 35 volleyballs the same week.
How many balls did she sell in all?

Add 47 + 35. Add the ones.	Regroup if you need to.	Add the tens.
$7 + 5 = 12$	12 ones = 1 ten 2 ones	$1 + 4 + 3 = 8$
[ ] 4 7 + 3 5	[1] 4 7 + 3 5 = 2	[1] 4 7 + 3 5 = 8 2

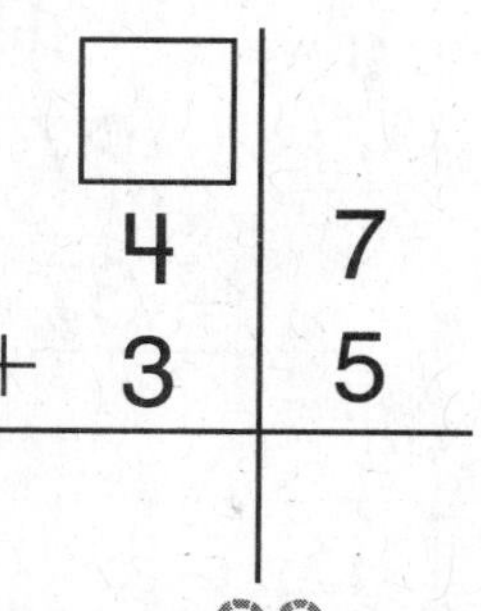

Eliza sold 82 balls in all.

Add. Regroup if you need to.

1. [1] 43 + 19 = 62
2. [ ] 56 + 18
3. [ ] 38 + 42
4. [ ] 23 + 45
5. 12 + 49
6. 81 + 17
7. 29 + 47
8. 51 + 38
9. 26 + 55
10. 42 + 23
11. 16 + 29
12. 38 + 37

NS2.2 Find the sum or difference of two whole numbers up to three digits long.

Name ________________________________

More 2-Digit Addition

Add. Regroup if you need to.

1. $\begin{array}{r} 1 \\ 28 \\ +\ 35 \\ \hline 63 \end{array}$	2. $\begin{array}{r} 47 \\ +\ 24 \\ \hline \end{array}$	3. $\begin{array}{r} 62 \\ +\ 31 \\ \hline \end{array}$	4. $\begin{array}{r} 53 \\ +\ 19 \\ \hline \end{array}$
5. $\begin{array}{r} 16 \\ +\ 51 \\ \hline \end{array}$	6. $\begin{array}{r} 78 \\ +\ 12 \\ \hline \end{array}$	7. $\begin{array}{r} 34 \\ +\ 52 \\ \hline \end{array}$	8. $\begin{array}{r} 27 \\ +\ 37 \\ \hline \end{array}$
9. $\begin{array}{r} 59 \\ +\ 31 \\ \hline \end{array}$	10. $\begin{array}{r} 22 \\ +\ 56 \\ \hline \end{array}$	11. $\begin{array}{r} 23 \\ +\ 45 \\ \hline \end{array}$	12. $\begin{array}{r} 44 \\ +\ 17 \\ \hline \end{array}$
13. $\begin{array}{r} 67 \\ +\ 29 \\ \hline \end{array}$	14. $\begin{array}{r} 18 \\ +\ 34 \\ \hline \end{array}$	15. $\begin{array}{r} 35 \\ +\ 16 \\ \hline \end{array}$	16. $\begin{array}{r} 78 \\ +\ 11 \\ \hline \end{array}$

Problem Solving

17. The boys' softball team has 34 balls and 28 mitts. The girls' softball team has 27 balls and 33 mitts. How many mitts do they have in all?

_____ mitts

Name ____________________

Problem Solving Skill: Use a Table

How many points did Lou and Becky score in all?

Points Scored This Season	
Player	**Number of Points**
Anna	26
Lou	37
Becky	23
Kevin	19

1. Look at the table. How many points did Lou score? 37 points

2. Look at the table. How many points did Becky score? 23 points

3. Add 37 and 23.

$$\begin{array}{r} ^{1} \\ 37 \\ +\ 23 \\ \hline 60 \end{array}$$

Lou and Becky scored  points in all.

Use the table to solve.

Number of Games Played	
Player	**Number of Points**
Marco	17
Jill	32
Patrick	27
Dakota	19

4. How many games did Marco and Patrick play in all?

____ games

5. How many games did Jill and Dakota play in all?

____ games

6. How many games did Jill and Patrick play in all?

____ games

NS2.0 Students estimate, calculate, and solve problems involving addition and subtraction of two- and three- digit numbers. AF1.2 Relate problem sitations to number sentences involving addition and subtraction.

Name ____________________

Problem Solving Skill: Use a Table

Use the table to solve.

Goals Scored This Season	
Player	Number of Goals
Terry	16
Jose	24
Darcy	27
Otto	18

1. How many goals did Jose and Darcy score in all?

   $\begin{array}{r} \overset{1}{2}4 \\ +\ 27 \\ \hline 51 \end{array}$

    goals

2. How many goals did Terry and Otto score in all?

   _____ goals

3. Which two players scored 45 points in all?

   ________ and ________

4. Which two players scored 40 goals in all? Add to check your answer.

   ________ and ________

Name ______________________________ **Week 11**

Spiral Review

Draw and label coins from greatest to least value. Find the total value.

1. 

______

Use the pictograph. 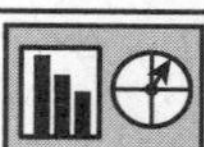

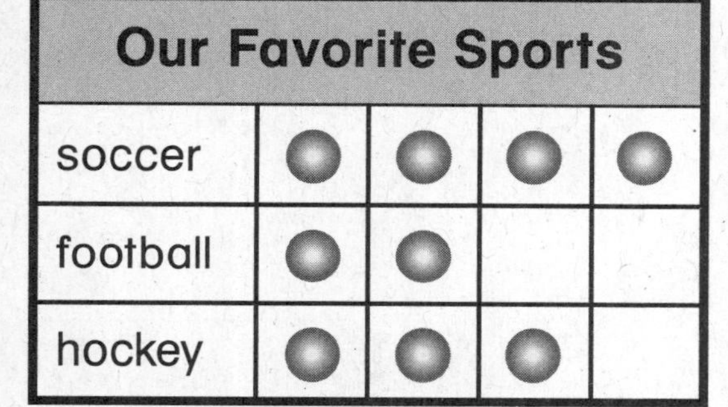

Our Favorite Sports				
soccer	●	●	●	●
football	●	●		
hockey	●	●	●	

Key: Each ● stands for 2 children.

2. How many children did not choose hockey?

______ children

Use a clock. 
Show the time.
Write the time.

3. 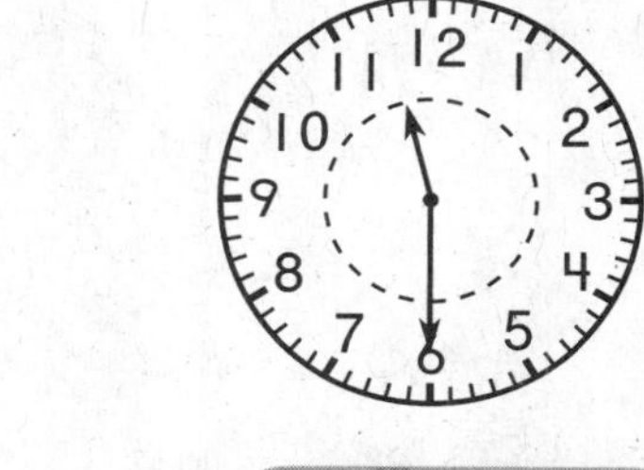

Find the difference. 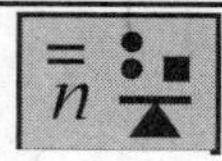
Complete the addition fact that can help.

4. $14 - 9 =$ ______

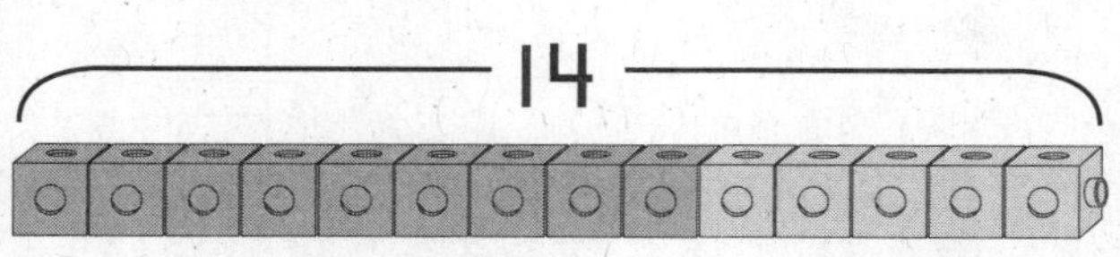

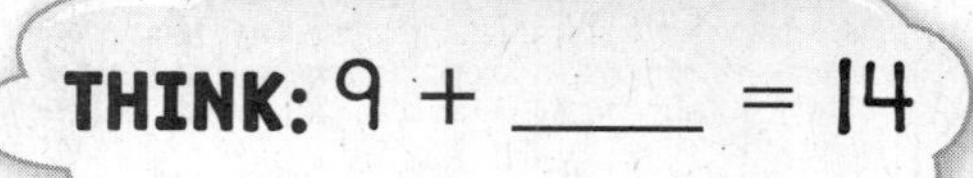
THINK: $9 +$ ______ $= 14$

5. $13 - 6 =$ ______

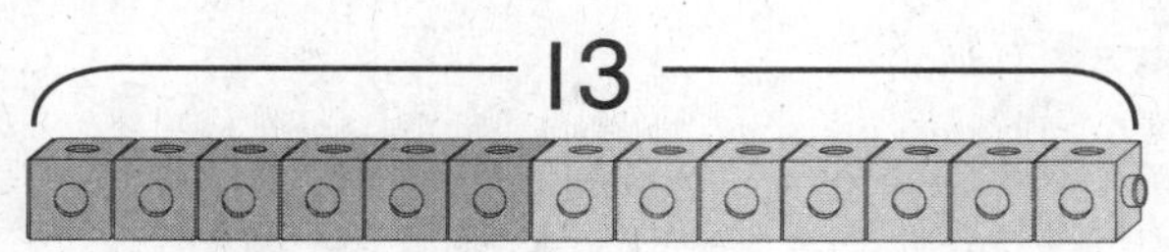

THINK: $6 +$ ______ $= 13$

Name____

Regrouping for Subtraction

Subtract 6 from 45.

Show 45 with [tens] [ones].
Count the ones.
Are there enough ones to subtract 6? no

So, you will have to **regroup**.

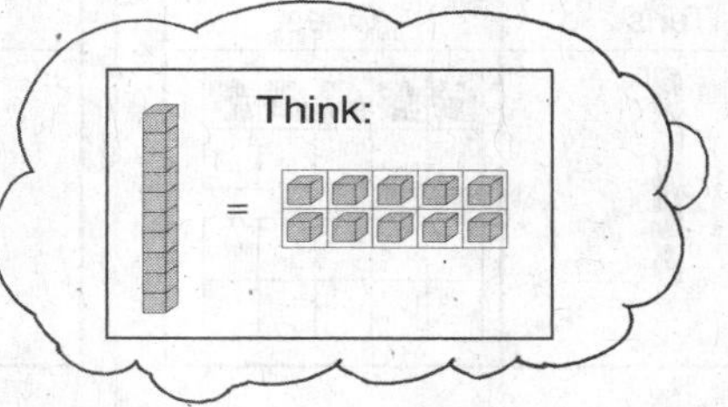

Trade 1 ten for 10 ones.

Subtract 6 ones from 15 ones.
How many ones are left? 9 ones

Subtract 0 tens from 3 tens.
How many tens are left? 3 tens

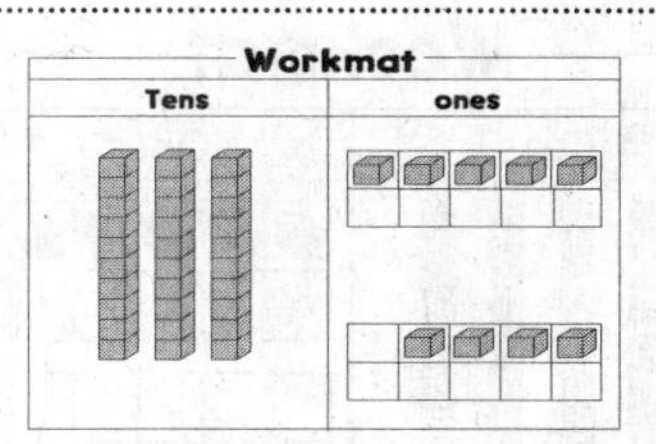

The difference is 39

3 tens 9 ones is the same as 39.

Use Workmat 11 and [tens] [ones]. Regroup if you need to.
Write how many **tens** and **ones**. Write the difference.

1. Subtract 5 from 34.

____tens ____ ones

______

2. Subtract 8 from 47.

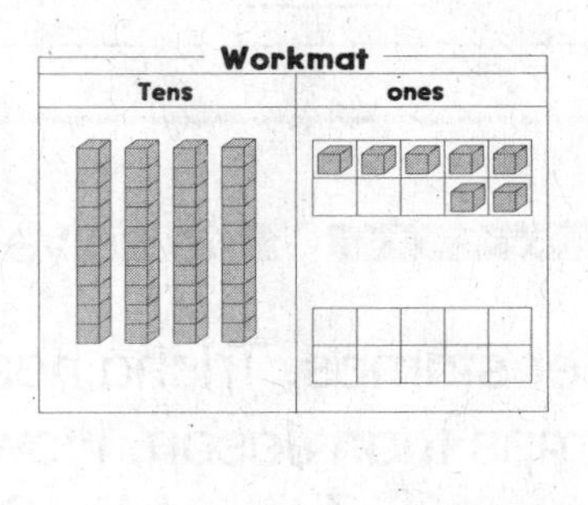

____tens ____ ones

______

3. Subtract 7 from 29.

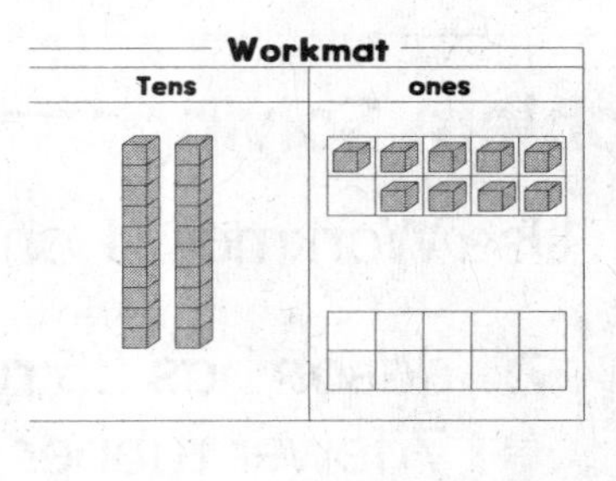

____tens ____ ones

______

AF 1.0 Students model, represent, and interpret number relationships to create and solve problems involving addition and subtraction. **NS 2.0** Students estimate, calculate, and solve problems involving addition and subtraction of two- and three-digit numbers. **MR 1.2** Use tools, such as manipulatives or sketches, to model problems.

Name____

Lesson 7.2

Hands On: Regrouping for Subtraction

Use Workmat 11 and [tens block] [ones block]. Regroup if you need to.
Write how many tens and ones. Write the difference.

1. Subtract 6 from 22.

Workmat — Tens | Ones

__1__ tens __6__ ones

__16__

2. Subtract 4 from 15.

Workmat — Tens | Ones

_____ tens _____ ones

_____

3. Subtract 7 from 36.

Workmat — Tens | Ones

_____ tens _____ ones

_____

4. Subtract 8 from 45.

Workmat — Tens | Ones

_____ tens _____ ones

_____

5. Subtract 5 from 52.

Workmat — Tens | Ones

_____ tens _____ ones

_____

6. Subtract 3 from 31.

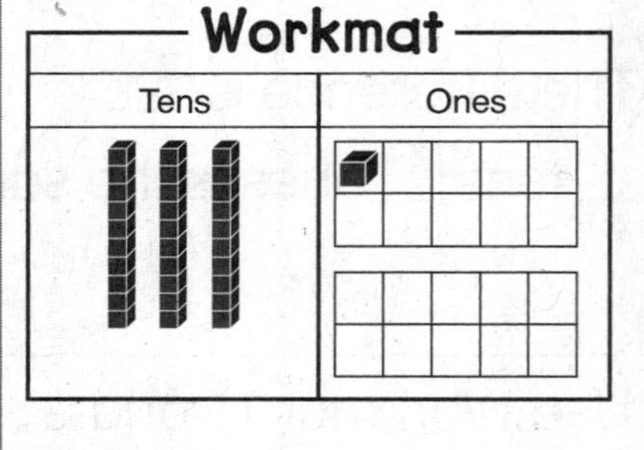

_____ tens _____ ones

_____

Problem Solving

Use Workmat 11 and [tens block] [ones block] to solve.

7. Jason has 25 rubber stamps. Trisha has 7 fewer rubber stamps than Jason. How many rubber stamps does Trisha have?

_____ rubber stamps

rubber stamp

Name ____

Model 2-Digit Subtraction

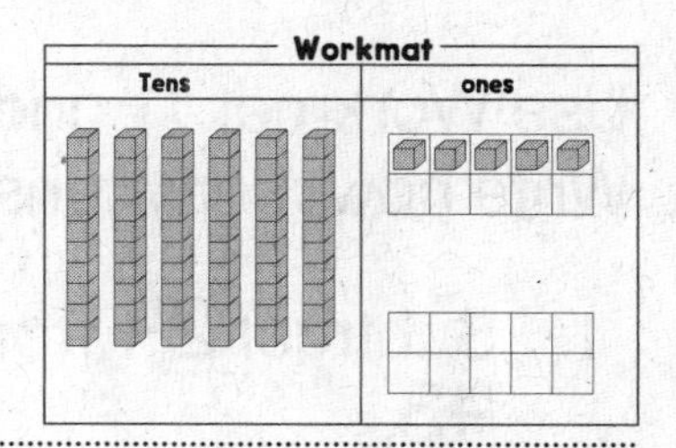

Subtract 37 from 65.
Show 65 with [tens block] [ones cube].
Are there enough ones to subtract 7? no

So, you will have to regroup.

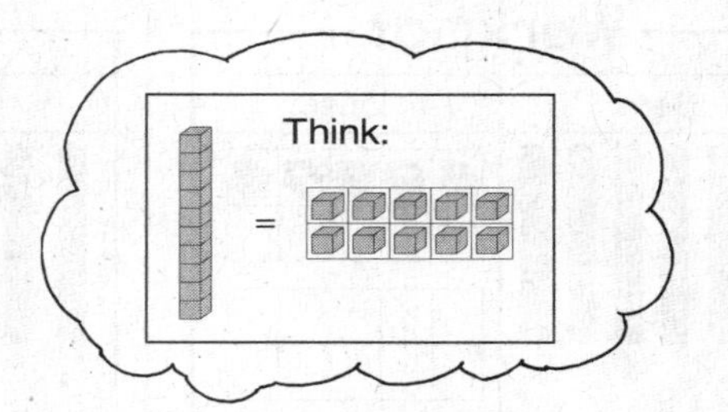

Trade 1 ten for 10 ones.

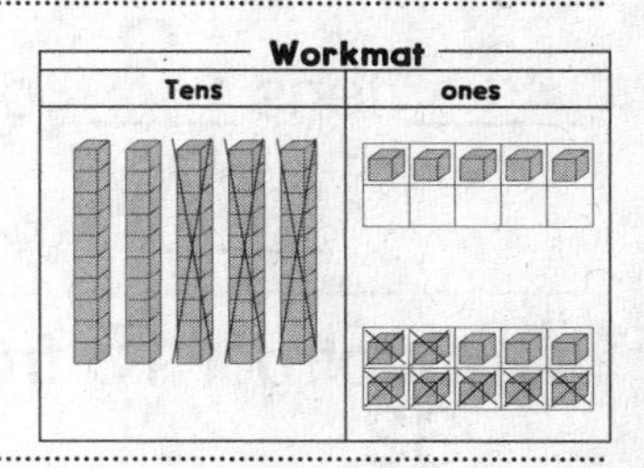

Subtract 7 ones from 15 ones.
How many ones are left? 8
Subtract 3 tens from 5 tens.

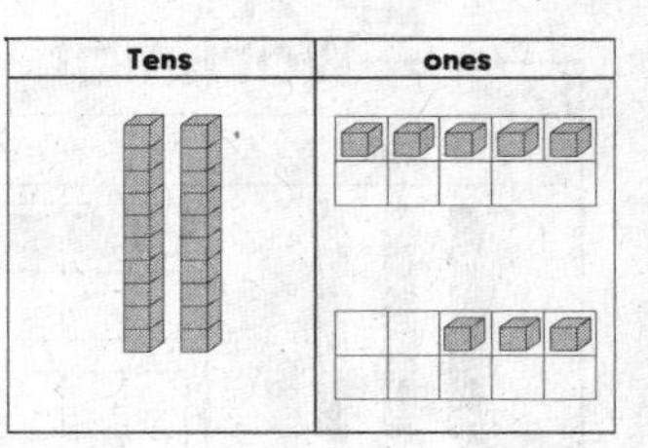

How many tens are left? 2 tens
The difference is 28.
2 tens 8 ones is the same as 28.

Use Workmat 11 and [tens block] [ones cube]. Regroup if you need to.
Write how many **tens** and **ones**. Write the difference.

1. Subtract 34 from 41.

____ tens ____ ones

______

2. Subtract 18 from 43.

____ tens ____ ones

______

3. Subtract 19 from 55.

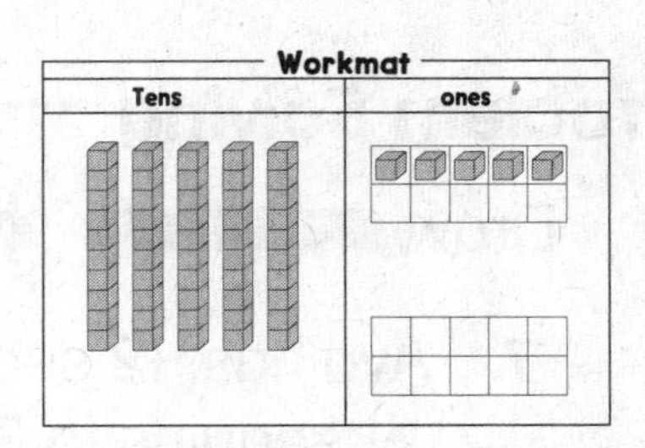

____ tens ____ ones

______

AF 1.0 Students model, represent, and interpret number relationships to create and solve problems involving addition and subtraction. NS 2.2 Find the sum or difference of two whole numbers up to three digits long. MR 1.2 Use tools, such as manipulatives or sketches, to model problems.

Name ________________________

Hands On: Model 2-Digit Subtraction

Use Workmat 11 and [tens block] [ones block]. Regroup if you need to.
Write how many tens and ones. Write the difference.

1. Subtract 24 from 43.

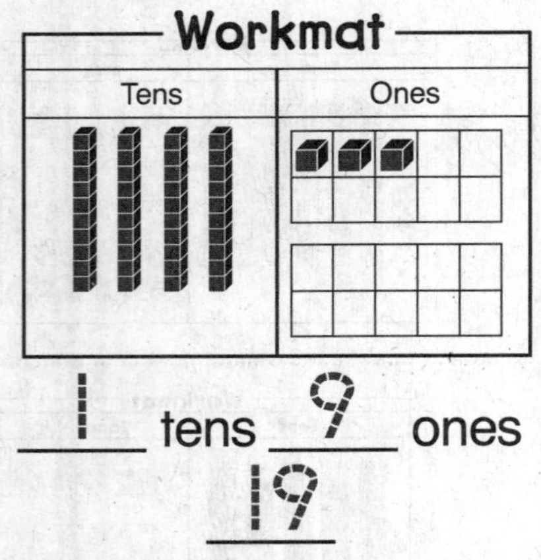

1 tens 9 ones

19

2. Subtract 17 from 65.

Workmat

Tens | Ones

_____ tens _____ ones

_____

3. Subtract 27 from 56.

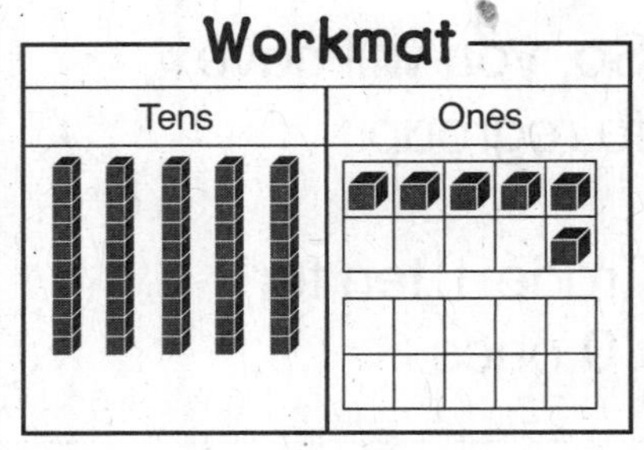

_____ tens _____ ones

_____

4. Subtract 15 from 33.

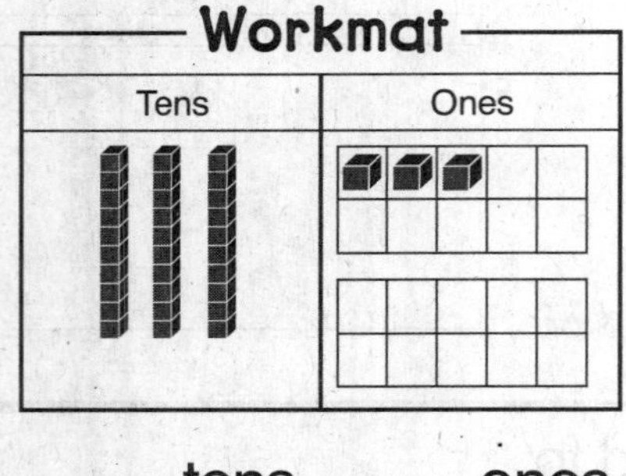

_____ tens _____ ones

_____

5. Subtract 46 from 68.

Workmat

Tens | Ones

_____ tens _____ ones

_____

6. Subtract 19 from 47.

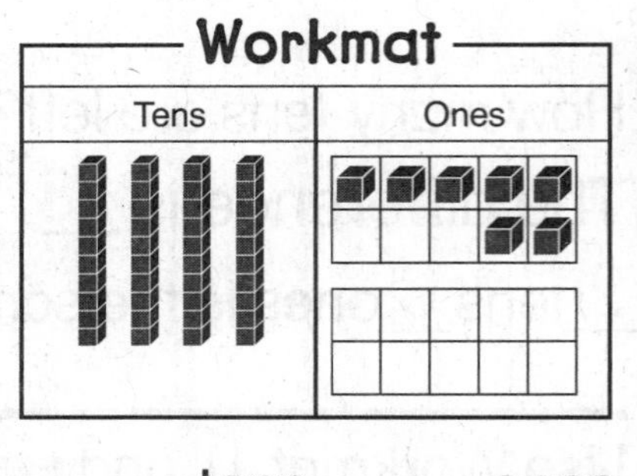

_____ tens _____ ones

_____

Problem Solving

Draw [tens block] [ones block] to solve.

7. Ava has 42 sequins. She gives 14 sequins to Beth. How many sequins does Ava have left?

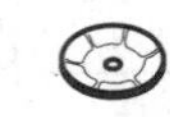

sequins

_____ sequins

Name ____

Lesson 7.4

Problem Solving Strategy: Make a Model

Dan's classmates make 31 origami figures. They put 15 figures on a table. They put the rest on a shelf. How many figures are on the shelf?

origami figure

Read to Understand

1. Circle the information.
Then underline the question.

Plan

2. How will you solve the problem?

make a model

Solve

3. Make a model for 31 figures.	Regroup if you need to.	Subtract the ones.	Subtract the tens.
Workmat: Tens, ones	Workmat: Tens, ones	Workmat: Tens, ones	Workmat: Tens, ones

How many tens are there? 1 ten

How many ones are there? 6 ones

How many figures are on the shelf? 16 origami figures

Check

4. Does your answer make sense? Yes. My model shows 16

Use Workmat 11 and [tens block] [ones cube] to make a model.
Write how many.

5. There are 52 buttons in a bucket.
35 of them are blue. The rest are green.
How many green buttons are in the bucket? ______ buttons

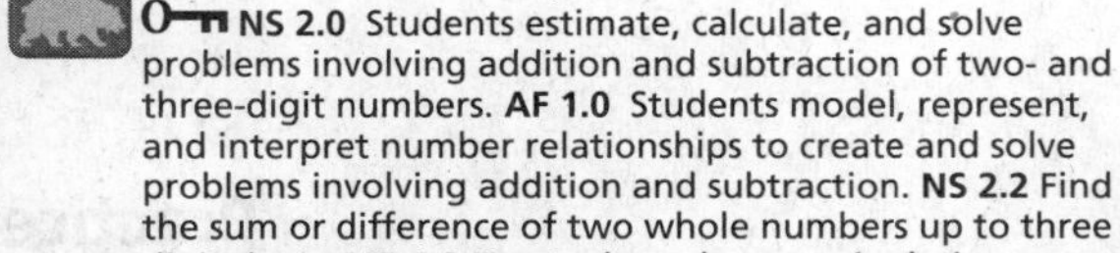

NS 2.0 Students estimate, calculate, and solve problems involving addition and subtraction of two- and three-digit numbers. **AF 1.0** Students model, represent, and interpret number relationships to create and solve problems involving addition and subtraction. **NS 2.2** Find the sum or difference of two whole numbers up to three digits long. **MR 1.2** Use tools, such as manipulatives or sketches, to model problems.

Name____ Lesson 7.4

Strategy • Make a Model

Problem Solving Strategy Practice

Use Workmat 1 1 and [tens block] [ones block] to make a model.
Write how many.

1. Juan has 52 moon shapes. He has 38 heart shapes. How many more moon shapes does he have than heart shapes? **moon shape**	____ more moon shapes
2. There are 45 paintbrushes in a cup. Dennis takes 28 paintbrushes out of the cup. How many paintbrushes are left in the cup? **paintbrush**	____ paintbrushes
3. Laura has 47 stickers. Chad has 9 fewer stickers than Laura. How many stickers does Chad have? Go You! **sticker**	____ stickers
4. There are 37 pieces of red yarn and some pieces of blue yarn in a box. There are 61 pieces of yarn altogether. How many pieces of blue yarn are in the box? **yarn**	____ pieces of blue yarn

Spiral Review

Find the total value.

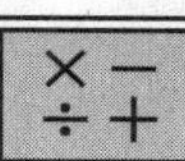

1.      

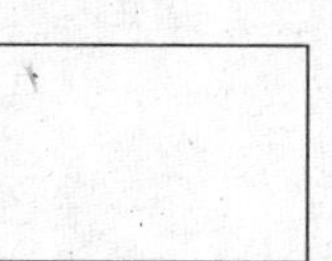

______, ______, ______, ______, ______, ______

Use the line plot.

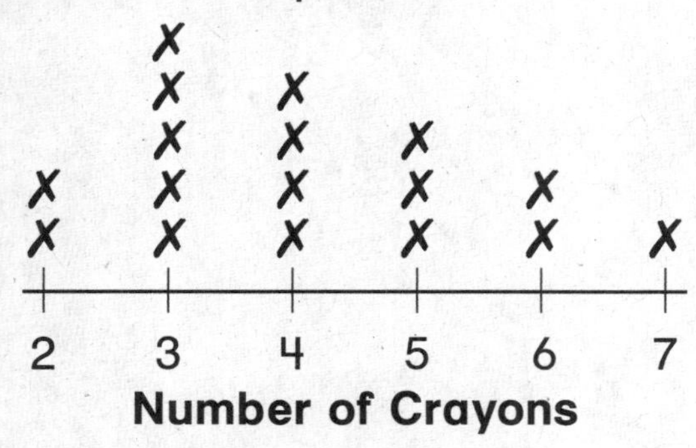

2. What number of crayons do the most children have?

______ crayons

Order 3 pieces of chalk from longest to shortest. Draw them.

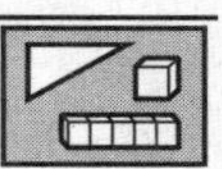

3. longest

4.

5. shortest

Use the table to solve.

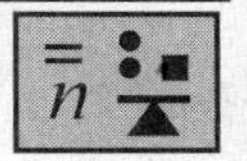

Flowers in a Garden	
Flower	**Number**
pansy	28
iris	19
rose	41
lily	16

6. How many more roses than irises are in the garden?

______ more roses

Name ____________________

Model and Record 2-Digit Subtraction

Subtract. 54
 − 5

Show 54. Are there enough ones to subtract 5? no

Workmat

Tens	ones

	tens	ones
	5	4
−		5

Regroup 1 ten as 10 ones.

Write the new number of tens and ones.

Workmat

Tens	ones

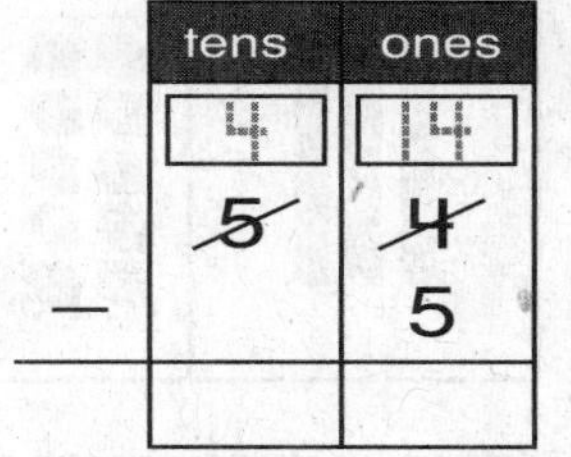

Subtract 5 ones from 14 ones.

How many ones are left? 9 ones
Write that number in the ones place.

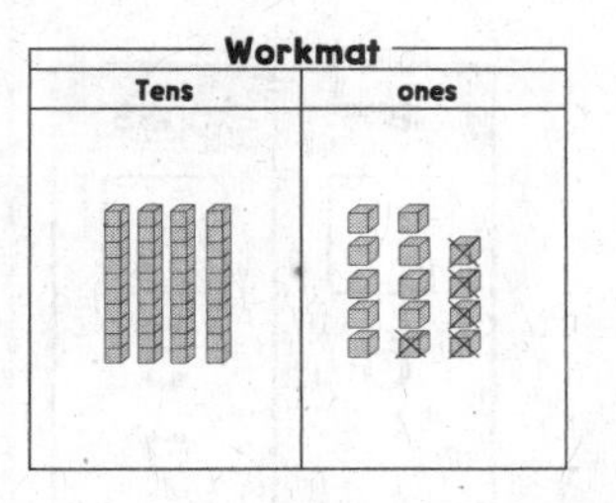

	tens	ones
	4	14
	~~5~~	~~4~~
−		5
		9

Subtract 0 tens from 4 tens

How many tens are left? 4 tens
Write that number in the tens place.

What is the difference? 49

Workmat

Tens	ones

	tens	ones
	4	14
	~~5~~	~~4~~
−		5
	4	9

Use Workmat 3 and ▭ ▫. Draw the regrouping if you need to. Write the difference.

1.

Workmat

Tens	ones

	tens	ones
	4	3
−		6

2.

Workmat

Tens	ones

	tens	ones
	3	1
−		7

Name______________________________

Lesson 7.5

Hands On: Model and Record 2-Digit Addition

Use Workmat 3 and [base-ten blocks].
Draw the regrouping if you need to. Write the difference.

1.

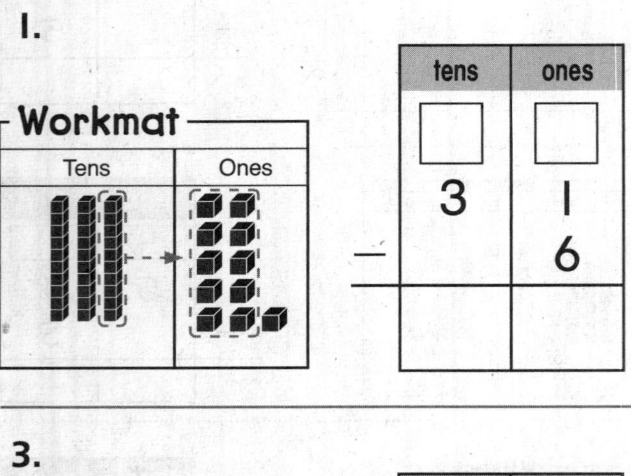

	tens	ones
	3	1
−		6

2.

Workmat

Tens	Ones

	tens	ones
	2	3
−		5

3.

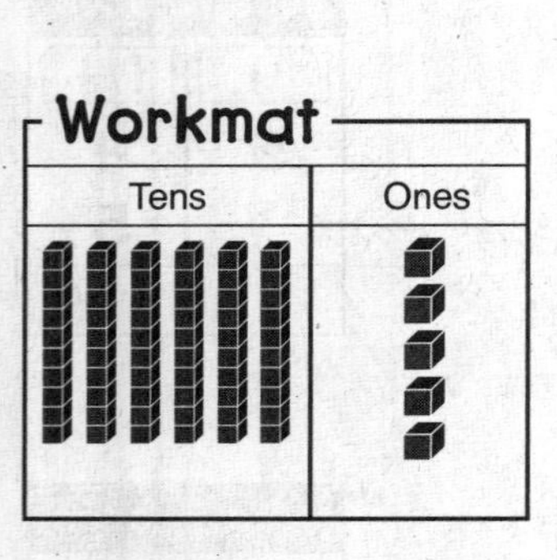

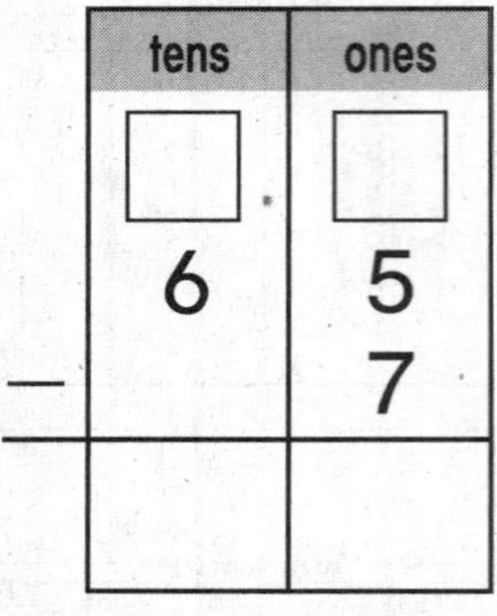

	tens	ones
	6	5
−		7

4.

Workmat

Tens	Ones

	tens	ones
	4	8
−		9

Problem Solving

Use Workmat 3 and [base-ten blocks] if you need to.

5. Mr. Thome has 24 paintbrushes. He gives 3 paintbrushes to one group of students and 5 paintbrushes to another group. How many paintbrushes does he have left?

paintbrush

______ paintbrushes

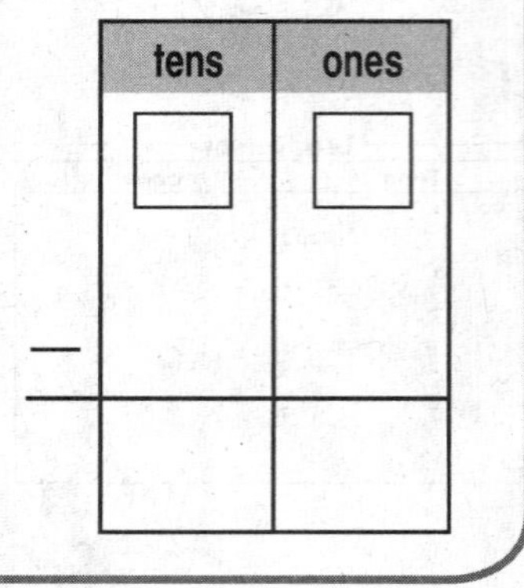

Name ______________________________ **Week 13**

Spiral Review

Circle coins to make \$1.00.
Cross out the coins you do not use.

1.

Use the bar graph.

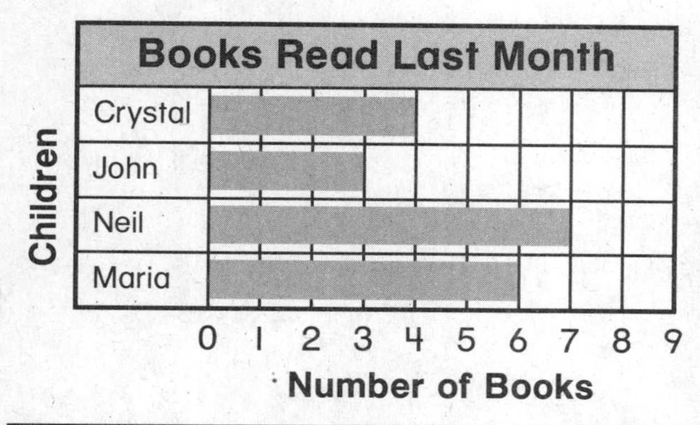

2. How many books did the children read in all?

_____ books

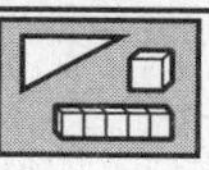

Cross out the objects that do not belong. Circle the objects that have the same figure. Name the solid figures you circled.

3.

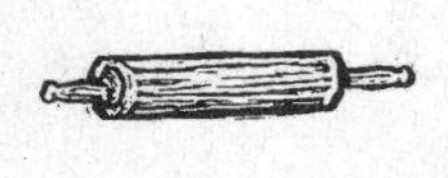 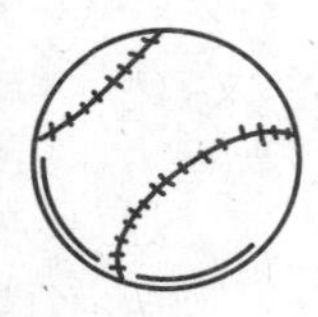  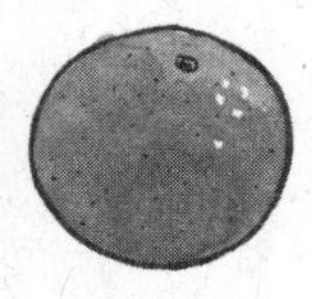

____

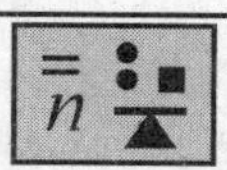

Use the table to solve.

Points Scored in the Tournament	
Team	**Number of Points**
Wildcats	32
Wolves	28
Sparrows	24
Eagles	36

4. How many more points did the Wildcats score than the Sparrows?

_____ more points

Name____

Estimate Differences

Write > or < to make the sentence true.

90 − 48 ◯ 40 | 90 − 52 ◯ 40

Use a difference you know to help you estimate.

90 − 50 = 40

When you subtract a smaller number, the difference will be greater.

48 is less than 50.

So, 90 − 48 (>) 40.

When you subtract a greater number, the diffference will be smaller.

52 is greater than 50.

So, 90 − 52 (<) 40.

Look at the difference in the box.
Write > or < to make each sentence true.

1. 50 − 40 = 10

50 − 38 (>) 10

50 − 41 ◯ 10

2. 60 − 10 = 50

60 − 9 ◯ 50

60 − 13 ◯ 50

3. 80 − 50 = 30

80 − 47 ◯ 30

80 − 52 ◯ 30

4. 40 − 20 = 20

40 − 22 ◯ 20

40 − 17 ◯ 20

NS 2.0 Students estimate, calculate, and solve problems involving addition and subtraction of two- and three-digit numbers. **NS 6.0** Students use estimation strategies in computation and problem solving that involve numbers that use the ones, tens, hundreds, and thousands places.
prob
diffe

Name ____________________

Estimate Differences

Look at the difference in the box.
Then write $>$ or $<$ to make each sentence true.

1. $80 - 40 = 40$

$80 - 37 > 40$

$80 - 42 \bigcirc 40$

2. $70 - 50 = 20$

$70 - 49 \bigcirc 20$

$70 - 51 \bigcirc 20$

3. $60 - 20 = 40$

$60 - 22 \bigcirc 40$

$60 - 17 \bigcirc 40$

4. $50 - 30 = 20$

$50 - 27 \bigcirc 20$

$50 - 34 \bigcirc 20$

5. $80 - 20 = 60$

$80 - 22 \bigcirc 60$

$80 - 19 \bigcirc 60$

6. $60 - 10 = 50$

$60 - 8 \bigcirc 50$

$60 - 11 \bigcirc 50$

Problem Solving

7. Write numbers in the blanks to make true sentences. Use the difference in the box to help.

$90 - 60 = 30$

$90 - \_\_\_\_ > 30$ | $90 - \_\_\_\_ < 30$

$90 - \_\_\_\_ > 30$ | $90 - \_\_\_\_ < 30$

$90 - \_\_\_\_ > 30$ | $90 - \_\_\_\_ < 30$

Spiral Review

Write how many tens and ones.
Use [tens rod] if you need to.

1. 33

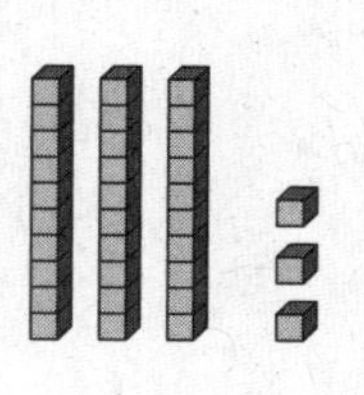

____ tens ____ ones

____ tens ____ ones

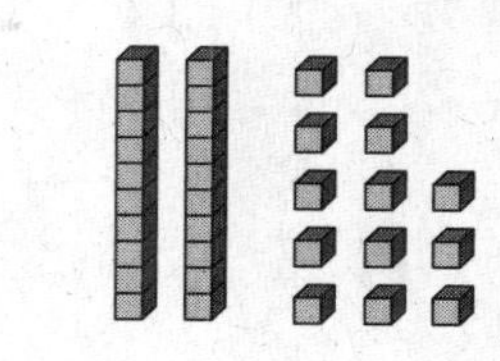

____ tens ____ ones

Find the pattern. Complete the table to solve.

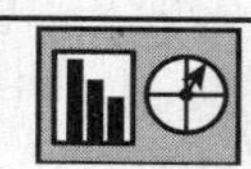

2. How many wheels are on 5 skateboards?

number of skateboards	1	2	3	4	5
number of wheels	4	8	12		

There are ____ wheels on each skateboard.

There are ____ wheels on 5 skateboards.

Complete the table to solve.

3. Daniel is making three cubes. How many squares does he need to make these three figures?

________________			
number of cubes	1	2	3
number of squares			

Daniel needs ____ squares.

Draw a picture to solve. Write the number sentence.

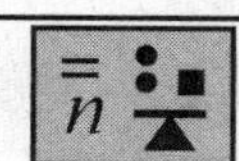

4. Becky sees 3 green butterflies and 4 purple butterflies. How many butterflies does Becky see altogether?

____ butterflies

Name ____________________

Lesson 8.7

Problem Solving Skill: Use a Table

How many more medium robots than extra-large robots are there?

Robot Toys in Store	
Size	**Number of Robots**
small	58
medium	41
large	25
extra-large	17

1. What are you asked to find out?

How many more medium than extra-large robots there are.

2. Look at the table.

How many medium robots are there? 41 medium robots

How many extra-large robots are there? 17 extra-large robots

3. Subtract to compare.

$$\begin{array}{r} \overset{3\ 11}{41} \\ -\ 17 \\ \hline 24 \end{array}$$

There are 24 more medium robots than extra-large robots.

Use the table to solve.

Blocks in Towers	
Child	**Number of Blocks**
Aidan	26
Stella	42
Carlos	22
Hakeem	38

5. How many more blocks are in Stella's tower than in Aidan's tower ?

_____ more blocks

6. How many blocks are in Stella's and Hakeem's towers together?

_____ blocks

NS 2.0 Students estimate, calculate, and solve problems involving addition and subtraction of two - and three-digit numbers. AF 1.3 Solve addition and subtraction problems by using data from simple charts, picture graphs, and number sentences. AF 1.2 Relate problem situations to number sentences involving addition and subtraction

Name________________________________

Lesson 8.7

Problem Solving Skill: Use a Table

Use the table to answer the questions.

Shirts in a Store	
Color	**Number of Shirts**
blue	45
green	54
yellow	27
red	36

1. How many more blue shirts than yellow shirts are in the store?

   ______ more blue shirts

2. The green shirts and red shirts are all on one shelf. How many shirts are on that shelf?

   ______ shirts

3. The store sells 18 blue shirts. How many blue shirts are left in the store?

   ______ blue shirts

4. How many more green shirts than red shirts are in the store?

   ______ more green shirts

Name

Lesson 8.8

Mental Math: Find Differences

You can use mental math to find differences.

$65 - 36 = ?$

First, look at the number you are subtracting. What number can you add to make a ten?

Think: $36 + ? = 40.$

Think: $6 + 4 = 10.$

You can add 4 to make a ten.

Add 4 to both numbers.

$65 - 36 = ?$

$+4 \quad +4$

$69 - 40 = ?$

Now you can use mental math Count back by tens.

69, 59, 49, 39, 29

$69 - 40 = 29$

So, $65 - 36 = 29$

	Add the same number to both numbers.	Subtract.
1. $73 - 18 = ?$	73 − 18 +___ +___ ___ − ___	___ − ___ = ___ So, 73 − 18 = ___
2. $64 - 47 = ?$	64 − 47 +___ +___ ___ − ___	___ − ___ = ___ So, 64 − 47 = ___

NS 2.3 Use mental arithmetic to find the sum or difference of two two-digit numbers. **NS 2.2** Find the sum or difference of two whole numbers up to three digits long.

Name ______________________________

Mental Math: Find Differences

	Add the same number to both numbers.	Subtract.
1. 64 − 36 = ?	64 − 36 + 4 + 4 68 − 40	68 − 40 = 28 So, 64 − 36 = 28
2. 85 − 48 = ?	85 − 48 + ___ + ___ ___ − ___	___ − ___ = ___ So, 85 − 48 = ___
3. 51 − 17 = ?	51 − 17 + ___ + ___ ___ − ___	___ − ___ = ___ So, 51 − 17 = ___

Problem Solving

Use mental math to solve.

4. Isak has 28 trading cards. He needs 42 trading cards to make a set. How many more trading cards does he need to make a set?

   ______ more trading cards

trading card

Name ______________________________ **Week 15**

Spiral Review

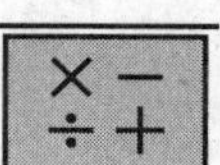

Write the fraction for the shaded part of the whole.

1.

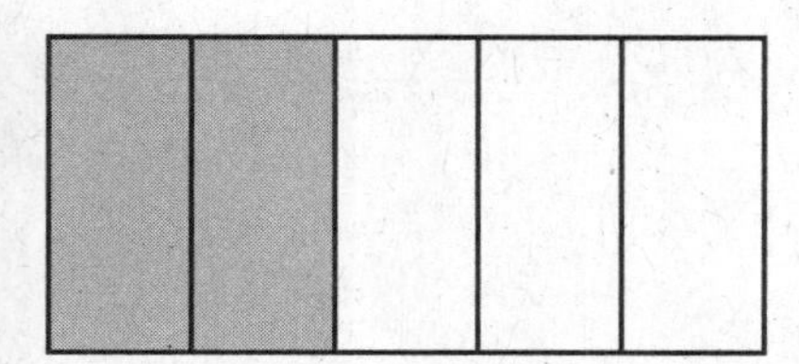

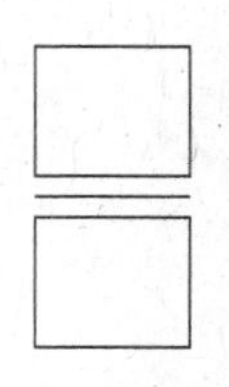

2.

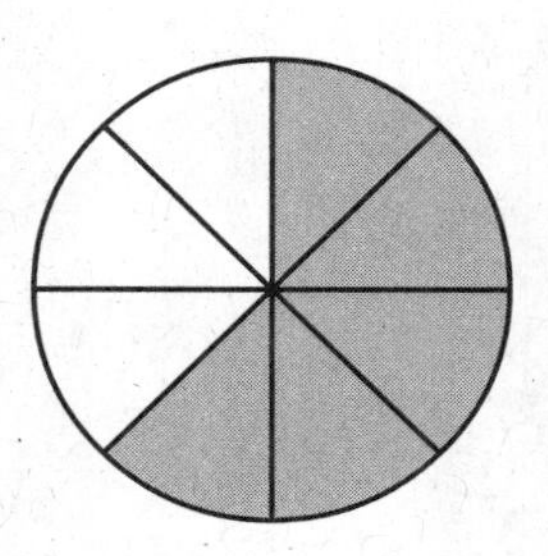

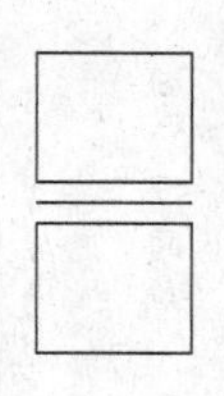

Use the picture to compete the table.
Then shade bars in the graph to show the data.

3.

Cars	
Color	Number
black	
gray	
white	

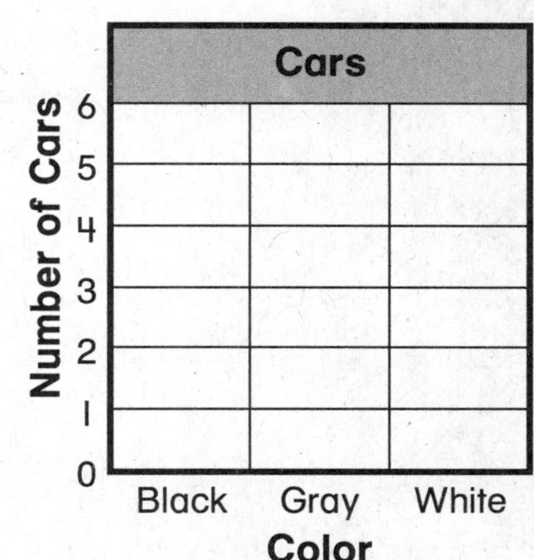

Color the figures.

4. Color the circles green.

5. Color the squares blue.

6. Color the triangles red.

7. Color the rectangles yellow.

Decide which operation to use.
Then write the number sentence. Solve.

8. Katie catches 16 lightning bugs in a jar. Then 8 of the lightning bugs get out. How many lightning bugs are left in the jar?

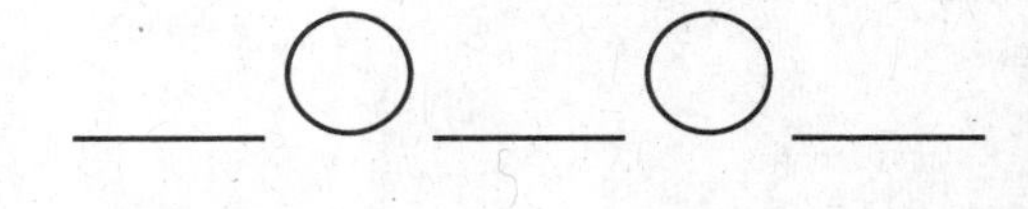

_____ lightning bugs

Name ____________________

Lesson 9.2

Take a Survey on a Tally Chart

When you take a **survey**, you ask many people the same question.

Show the answers on a **tally chart**.

Make one **tally mark** for each answer.

Favorite School Subjects	
Subject	**Tally**
math	II
science	~~IIII~~
reading	III

Two people said that math is their favorite subject.

---

1. Take a survey. Ask 10 classmates which season is their favorite. Use tally marks to show their answers.

Favorite Season	
Season	**Tally**
winter	
spring	
summer	
fall	

2. How many classmates chose summer as their favorite? ______ children

3. Did more classmates choose spring or fall? ____________

4. The fewest classmates chose ____________ as their favorite season.

SDAP1.0 – Students collect numerical data and record, organize, display, and interpret the data on bar graphs and other representations. **SDAP1.1** – Record numerical data in systematic ways, keeping track of what has been counted. **SDAP1.2** – Represent the same data set in more than one way (e.g., bar graphs and charts with tallies). **SDAP1.4** – Ask and answer simple questions related to data representations.

Name ____________________

Take a Survey on a Tally Chart

1. Take a survey. Ask 10 classmates which figure is their favorite. Use tally marks to show their answers.

Our Favorite Figures

Figure	Tally
square	
circle	
triangle	

2. Draw a stick-figure symbol in the picture graph to show the information in the tally chart.

Our Favorite Figures

square								
circle								
triangle								

Key: Each stick-figure symbol stands for 1 child.

Problem Solving

3. Mason wants to ask 12 classmates which sandwich is their favorite. Look at the results he has recorded. How many more classmates does he need to ask?

_______ more classmates

Our Favorite Sandwiches

Sandwich	Tally
ham	|||
turkey	|
peanut butter	~~||||~~

Name

Lesson 9.5

Problem Solving Strategy: Make a Bar Graph

Alyssa took a survey of her friends' favorite colors. 3 friends chose green, 8 friends chose blue, and 4 friends chose purple. How can she make a bar graph to show this data?

Read to Understand

1. What information will you use?
the number of friends who chose each color

Plan

2. How will you solve the problem?
make a bar graph to show the data

Solve

3. Make the bar graph to show the information in the problem.

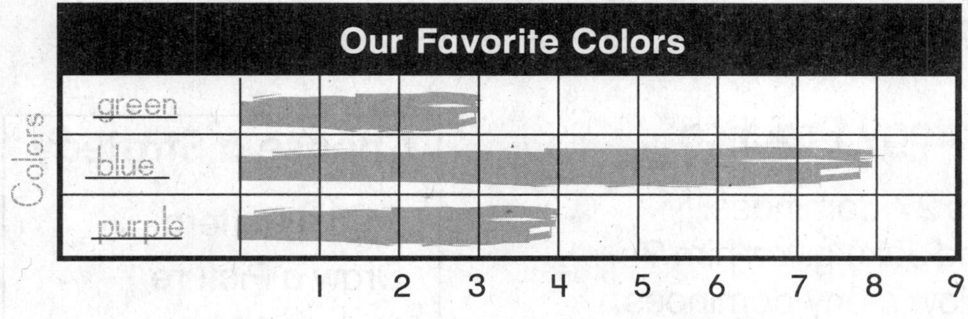

Check

4. Does your answer make sense?
Yes. My bar graph shows the correct data.

Use the bar graph to answer the question.

5. How many children did not choose blue? ______ children

Name____ **Lesson 9.5**

Strategy • Make a Bar Graph

Addie took a survey of her classmates' favorite pets. 8 classmates chose dogs, 4 chose cats, 6 chose fish, and 3 chose hamsters.

1. Complete the bar graph to show the information. Then write a title for the bar graph.

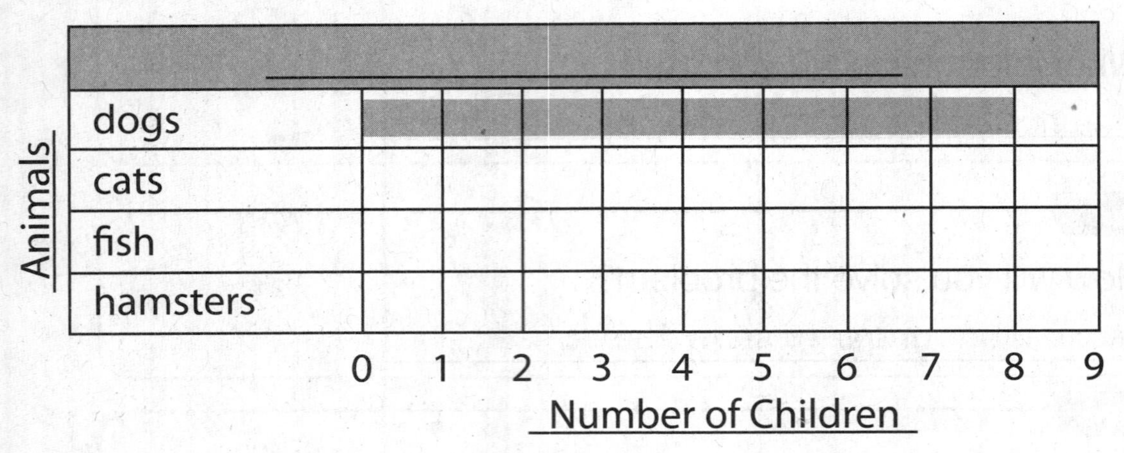

Use the bar graph to answer the question.

2. How many children did not choose dogs? ______ children

Mixed Strategy Practice

Choose a Strategy

- Find a Pattern
- Draw a Picture
- Make a Model

3. Sam has 27 dominoes. His friend Jake gives him 9 more. How many dominoes does Sam have now?

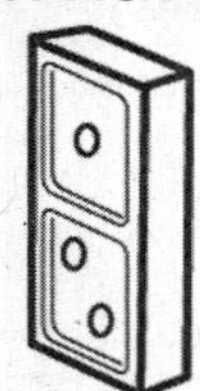

______ dominoes

4. The first number in a number pattern is 12. A rule for the pattern is **count by 3s**. What is the fourth number in the pattern?

______

Spiral Review

Write a fraction for the shaded part.

1.

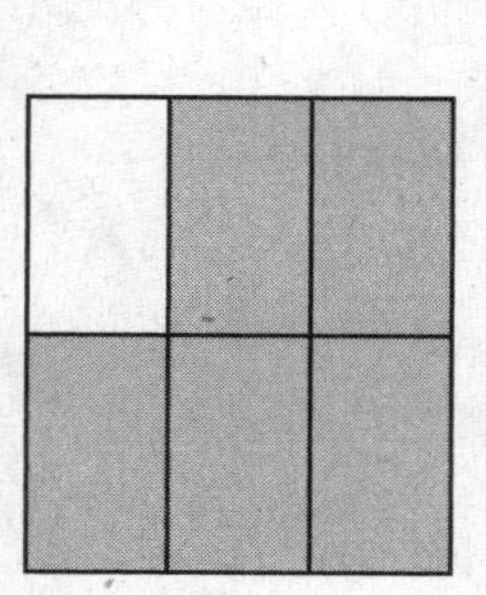

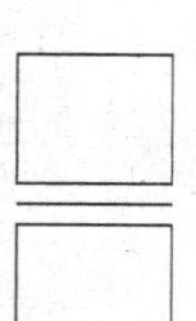

2.

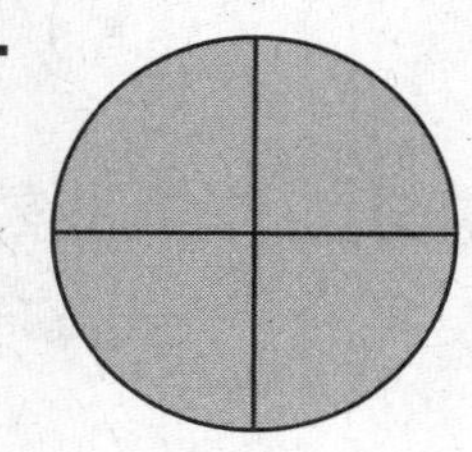

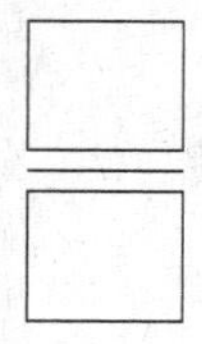

Use the line plot.

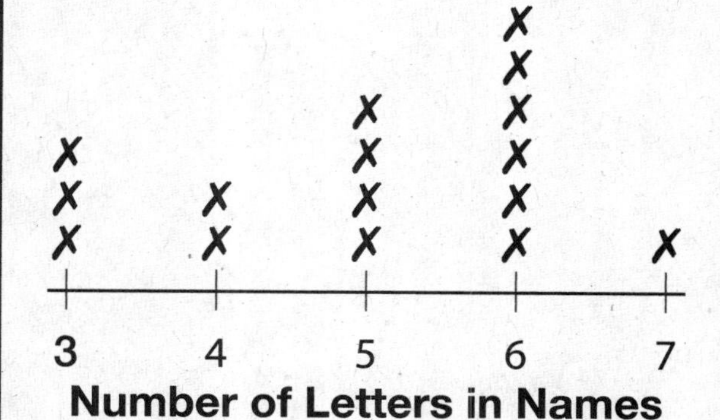

3. What is the mode of the data?

The mode of the data is _____.

Color the figures.

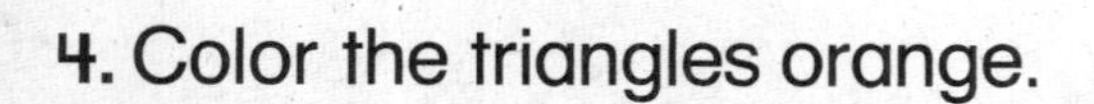

4. Color the triangles orange.

5. Color the squares purple.

6. Color the rectangles green.

7. Color the circles brown.

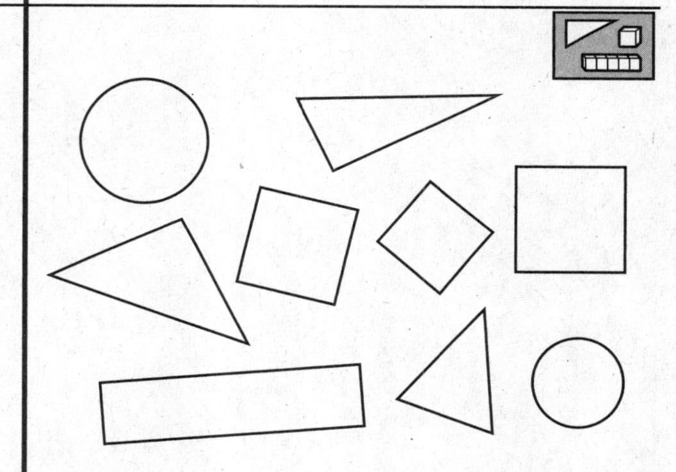

Find the sums.

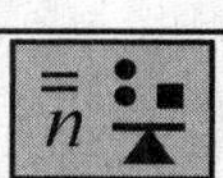

8.

$7 + 7 =$ ____

$7 + 8 =$ ____

9.

____ $= 6 + 0$

____ $= 1 + 0$

10.

$3 + 9 =$ ____

$9 + 3 =$ ____

Name ______________________ **Week 17**

Spiral Review

Subtract. Regroup if you need to.

1.	2.	3.	4.	5.
63 − 19	48 − 26	24 − 16	84 − 38	72 − 59

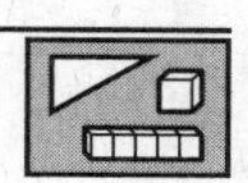

Skip-count. Write the missing numbers.
Write a rule for the pattern.

6.

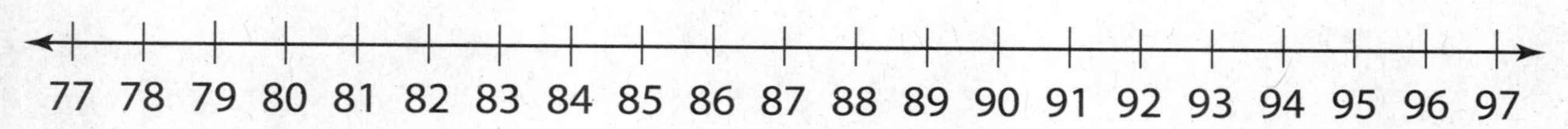

77, 79, 81, ____, ____   Rule: Count by ______.

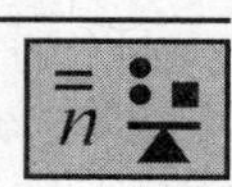

Use the solid figure.
Complete the table. Write how many.

7.

solid figure	faces	edges	vertices
pyramid	____	____	____

Complete the fact families.

8.

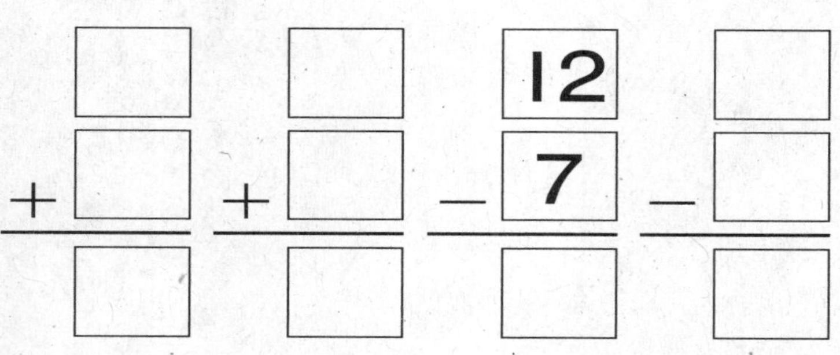

Name____

Lesson 10.4

Make the Same Amounts

You can show the same value in different ways.

You can show 60¢ with I half dollar and I dime.

=

You know that I half dollar is equal to 2 quarters.

=

You know that I dime is equal to 2 nickels.

So, you can also show 60¢ with 2 quarters and 2 nickels.

Use coins. Show the value in two ways.
Draw and label each coin.

1.

2.

Name____

Lesson 10.4

Hands On: Make the Same Amounts

Use coins. Show the value in two ways.

Draw and label each coin.

1.

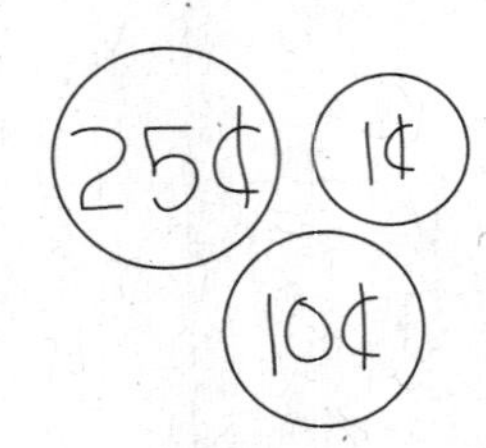

2.

3.

4.

Problem Solving

5. Mariel has three coins with a total value of 60¢. What coins could she have? Draw and label the coins.

Name ____________________

Lesson 10.5

Problem Solving Stategy: Make a List

Mia has 45¢. What are some different ways to use dimes and nickels to show how much money she has?

Read to Understand

1. What is being asked?
What are some ways to show 45¢ using dimes and nickels?

Plan

2. How do you solve the problem? make a list

Solve

3. The total value of the coins in each row should be 45¢.

Dimes	Nickels	Total Value
4	1	45¢
3	3	45¢
2	5	45¢

Check

4. Does your answer make sense?
Yes. The total value of each row is 45¢.

Make a list to solve.

5. Erin has 30¢. What are some different ways to use nickels and pennies to show how much money she has?

Nickels	Pennies	Total Value
____	____	____
____	____	____
____	____	____
____	____	____

NS5.1 Solve problems using combinations of coins and bills. NS5.0 Students model and solve problems by representing, adding, and subtracting amounts of money. NS5.2 Know and use the decimal notation and the dollar and cent symbols for money. MR2.0 Students solve problems and justify their reasoning.

Name_______________________________ **Lesson 10.5**

Strategy • Make a List

Problem Solving Strategy Practice

Make a list to solve.

1. Annie has 57¢. What are some different ways to show how much money she has using dimes, nickels, and pennies?

Dimes	Nickels	Pennies	Total Value
5	1	2	57¢

Mixed Strategy Practice

Choose a Strategy

- Make a List
- Find a Pattern
- Draw a Picture

Choose a strategy to solve.

2. Dora is trading nickels for pennies.
She trades 1 nickel for 5 pennies.
She has 5 nickels.
How many pennies should she get for 5 nickels?

number of nickels	1	2	3	4	5
number of pennies	5				

_______ pennies

3. Dean has 10 marbles. He gets 5 more.
How many marbles does he have now?

_______ marbles

Name________________________________

Compare Amounts

REMEMBER: count the coins from greatest to least value

To compare these two amounts, first find the total value of each group of coins

Count the coins.

25¢, 50¢, 55¢, 60¢, 65¢

Count the coins.

25¢, 35¢, 45¢, 55¢, 60¢

Then, compare the total values.
Write >, <, or =.

65¢ (>) 60¢

REMEMBER:
> means is greater than
< means is less than
= means is equal to

Write the total value of each group.
Then write >, <, or =.

1.

______ ◯ ______

2.

______ ◯ ______

NS 5.0 Students model and solve problems by representing, adding, and subtracting amounts of money. NS 5.2 Know and use the decimal notation and the dollar and cent symbols for money.

Name________________________________

Lesson 11.1

Compare Amounts

Write the total value of each group.
Then write >, <, or =.

1.

42¢ (>) 32¢

2.

_____ ( ) _____

3.

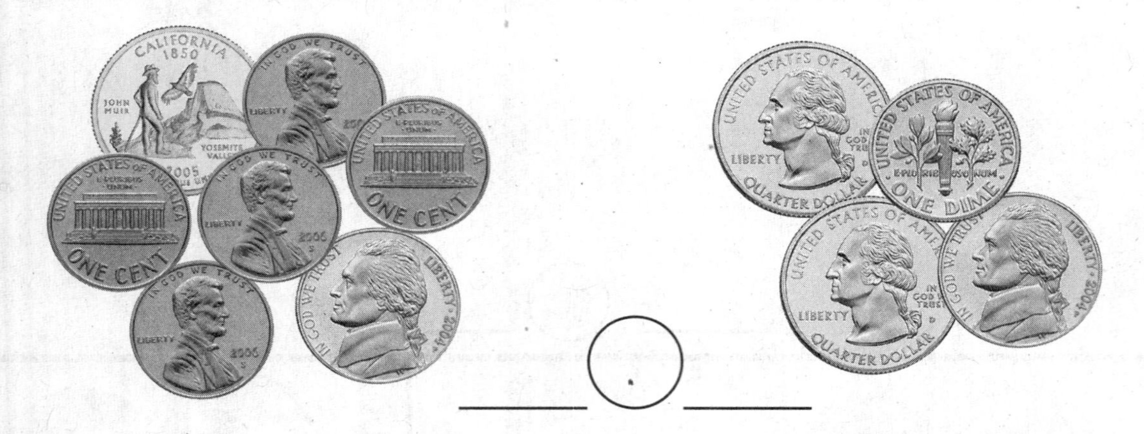

_____ ( ) _____

Problem Solving

4. Sandra buys a rose for 1 half dollar and 1 dime. She buys a daisy for 2 quarters and 1 nickel. Circle the flower that costs more money.

rose

daisy

rose daisy

Spiral Review

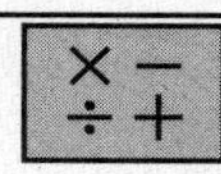

Write the number another way.

1. 2 hundreds 1 ten 7 ones

____

2. 942

____

3. five hundred forty-eight

____

4. 700 + 80 + 3

____

Use the line plot to answer the questions.

0	1	2	3	4
		X		
		X		
	X	X		
X	X	X		X
X	X	X	X	X

Number of Pets

5. Write a number sentence to find the range of the data.

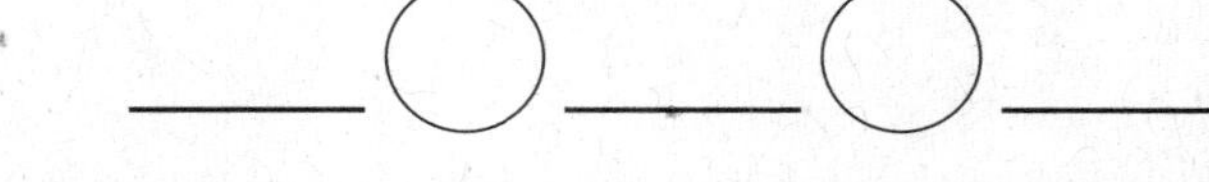

6. What is the mode of the data?

The mode of the data is ______.

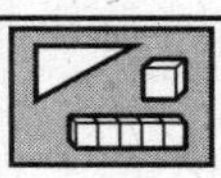

Use pattern blocks. Combine sides of figures to make a new figure. Trace the new figure.

7.

Draw a picture to solve.
Write the number sentence.

8. There are 4 ladybugs sitting on a branch. Then 7 more ladybugs come. How many ladybugs are there in all?

______ ladybugs

Name____

Problem Solving Strategy: Act it Out

Matt has 1 nickel, 2 quarters, and 2 dimes.
He buys a box of pencils.
Which coins, if any, are left over?

Read to Understand

1. Circle the information. Underline the question.

Plan

2. How will you solve the problem? act it out

Solve

3. Draw the coins. Write the total value. Then cross out coins to show the amount Matt spends.

75¢

Which coins are left? 1 quarter, 1 nickel.

Check

4. Does your answer make sense?
Yes. 75¢ − 45¢ = 30¢.

Use coins to act out the problem. Draw the coins. Write the total value. Then answer the question.

5. Pete has 1 half dollar and 3 dimes. He buys a toy for 80¢. Which coins, if any, are left over?

______________________ __________

6. Tammy has 3 quarters, 2 dimes, and 2 pennies. She buys a pen for 52¢. Which coins, if any are left over?

______________________ __________

NS5.0 — Students model and solve problems by representing, adding, and subtracting amounts of money. **NS5.1** — Solve problems using combinations of coins and bills. **NS5.2** — Know and use the decimal notation and the dollar and cent symbols for money.

Name____

Strategy • Act It Out

Problem Solving Strategy Practice

Use coins to act out a problem. Draw the coins. Write the total value. Then answer the question.

1. Donna has 1 quarter, 2 dimes, and 3 nickels. Does she have enough to buy the pen?

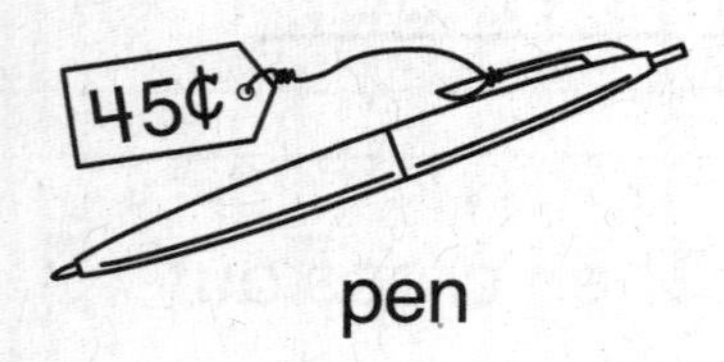

pen

_______ _______

2. Jay has 1 half dollar and 2 nickels. Does he have enough to buy the toy duck?

toy duck

_______ _______

Mixed Strategy Practice

Choose a strategy to solve.

Choose a Strategy
• Act It Out • Make a List • Draw a Picture

3. Jill has 2 quarters, 2 dimes, and 2 nickels. She gets 3 more nickels. How much money does she have?

____

4. Nate has 1 quarter and 1 dime. He wants to change his coins for nickels. How many nickels should he get?

_______ nickels

Name____________________

Lesson 11.3

Add and Subtract Money

Kara has 56¢. She gets 25¢ from her sister.
How much money does Kara have now?

Add when someone gets more money.

Reread the problem carefully.
Look for words that tell you to add or subtract.

Subtract when someone gives away or spends money.

Then, add or subtract the money amounts the same way you add and subtract other numbers.

```
  1
  56¢  ← the amount Kara starts with
+ 25¢  ← the amount her sister gives her
-----
  81¢  ← the amount Kara has now
```

Now Kara has 81¢.

Add or subtract to solve.

1. Gary has 61¢.
   He spends 17¢.

   Now Gary has ______.

2. Pat has 34¢.
   She finds 19¢.

   Now Pat has ______.

3. Beau has 53¢.
   He gives his sister 25¢.

   Now Beau has ______.

NS5.0 — Students model and solve problems by representing, adding, and subtracting amounts of money. NS5.2 — Know and use the decimal notation and the dollar and cent symbols for money.

Name ____________________

Add and Subtract Money

Add or subtract to solve.

1. Teri has 42¢. Her mom gives her 19¢.

   Now Teri has __________.

2. Noah has 75¢. He gives his brother 25¢.

   Now Noah has __________.

3. Jen has 38¢. She finds 15¢.

   Now Jen has __________.

4. Juan has 87¢. He spends 39¢.

   Now Juan has __________.

5. Todd has 94¢. He buys some juice for 56¢.

   Now Todd has __________.

Problem Solving

6. Marie takes 22¢ from her coin bank. Now she has 49¢ left. How much money did she start with?

   ____

Problem Solving Strategy: Predict and Test

Henry bought two items for 65¢.
One item was a bucket.
What was the other item?

Read to Understand

1. Circle the important information.
   Predict an answer to the problem. Prediction: boat

Plan

2. How will you solve the problem? predict and test

Solve

3. Write your prediction. Then test it to solve.

Predict	Test	Compare
boat 22¢	43¢ +22¢ = 65¢	65¢ (=) 65¢

Henry bought a bucket and a boat.

Check

4. Does your answer make sense?
   Yes. The bucket and the boat together cost exactly 65¢.

Make a prediction and check it to solve.

5. Ryan bought two items for 78¢.
   One item was a shovel.
   What was the other item?
   ____

6. Rob has 73¢. Which two items can he buy with exactly 73¢?
   ____

NS 5.0 Students model and solve problems by representing, adding, and subtracting amounts of money. NS 5.2 Know and use decimal notation and the dollar and cent symbols for money.

Name____

Strategy • Predict and Test

Problem Solving Strategy Practice

Make a prediction and check it to solve.

1. Dee has 68¢. Which two toys can she buy?

____

____

2. Terry has 91¢. Which two toys can he buy for exactly that amount?

____

____

Mixed Strategy Practice

Choose a strategy to solve.

Choose a Strategy
• Predict and Test • Make a Model • Draw a Picture

3. Diego has 7 nickels and 15 dimes. How many coins does Diego have in all?

_______ coins

4. Elsa has 32 pennies. Paul has 29 pennies. How many pennies do they have in all?

_______ pennies

Name____

One Dollar

one hundred cents is equal to one **dollar**.

100¢ = $1.00

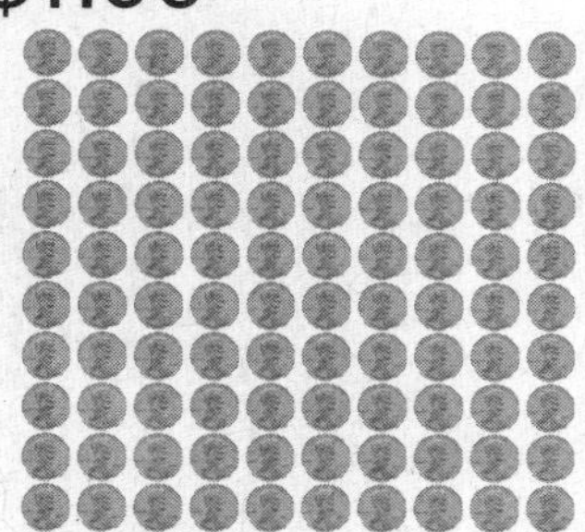

=

When you count to 1 dollar, you count to 100 cents.

25¢, 50¢, 75¢, $1.00

The **$** shows dollars. It is a **dollar sign**.

The **.** separates the dollars from the cents. It is a **decimal point**.

Use coins. Draw and label the coins to show $1.00.

1. dimes

$1.00

2. half dollars

______

3. nickels

______

NS 5.0 Students model and solve problems by representing, adding, and subtracting amounts of money. NS 5.2 Know and use the decimal notation and the dollar and cent symbols for money.

Name______________________________ **Lesson 11.5**

One Dollar

Circle coins to make $1.00.
Cross out the coins you do not use.

1.

2.

3.

Problem Solving

4. Keisha has $1.00 in her bag.
She has 2 of one kind of coin.
She has 1 of another kind of coin.
Draw the coins she has.

Spiral Review

Write how many equal parts there are. Then write if these parts are **halves**, **thirds**, or **fourths.**

1. 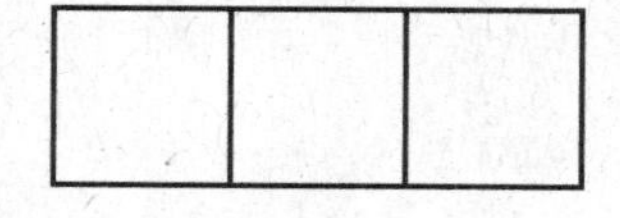

_____ equal parts

____

2. 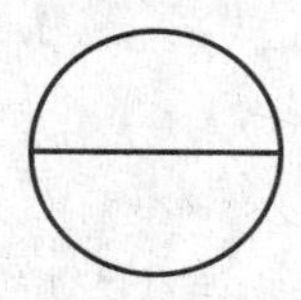

_____ equal parts

____

3. 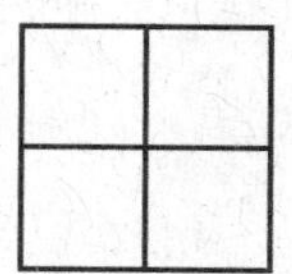

_____ equal parts

____

Use the line plot.

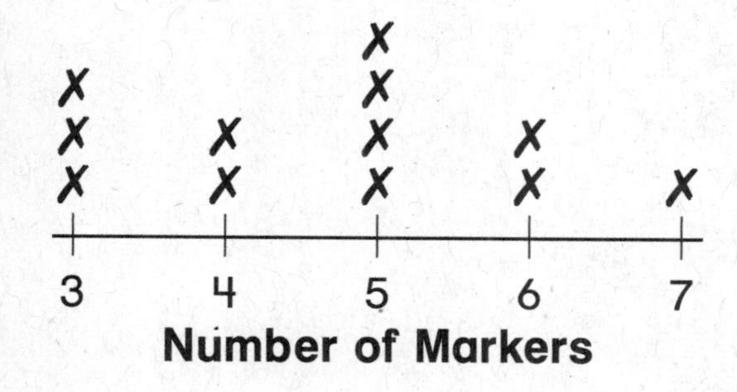

4. Write a number sentence to find the range of the data.

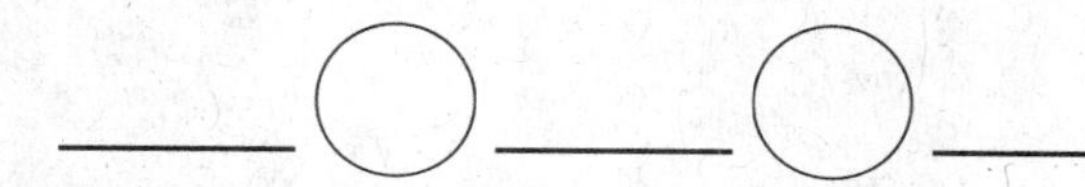

5. What is the mode of the data?

The mode of the data is _____.

Write a title to describe the group of plane figures.

6. __________

____

Write the sum.

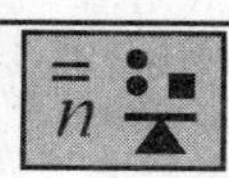

7. $6 + 7 + 4 =$ ____

8. $5 + 8 + 2 =$ ____

9. $7 + 1 + 7 =$ ____

10. $3 + 9 + 4 =$ ____

Name ____

Make Change to $1.00

When you pay for an item with more money than the cost, you get **change** back.

You have:	You buy:	Count on:
 50¢		 40¢, 50¢ The change is 11¢.

Find the total value of your coins.	Count on coins to find the change. Start at the price.	Count the change.
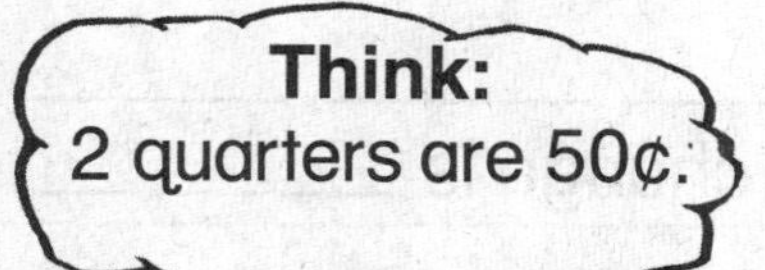 **Think:** 2 quarters are 50¢.	 39¢, 40¢, 50¢	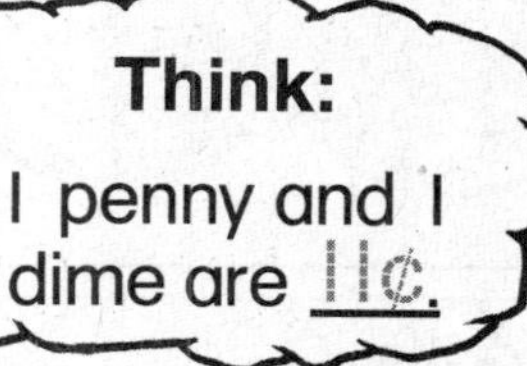 **Think:** 1 penny and 1 dime are 11¢.

So, the change is 11¢.

Count on from the price to make the change.

1. You have:	You buy:	Count on:
 ________		  ________, ________ Your change is ______.

2. You have:	You buy:	Count on:
 ________		  ________, ________ Your change is ______.

NS 5.0 Students model and solve problems by representing, adding, and subtracting amounts of money. NS 5.2 Know and use the decimal notation and the dollar and cent symbols for money.

Name ______________________

Lesson 11.6

Make Change to $1.00

Count on from the price to find the change.

1. You have

____

You buy

84¢

chalk

85¢, 90¢, $1.00

Your change is 16¢.

2. You have

____

You buy

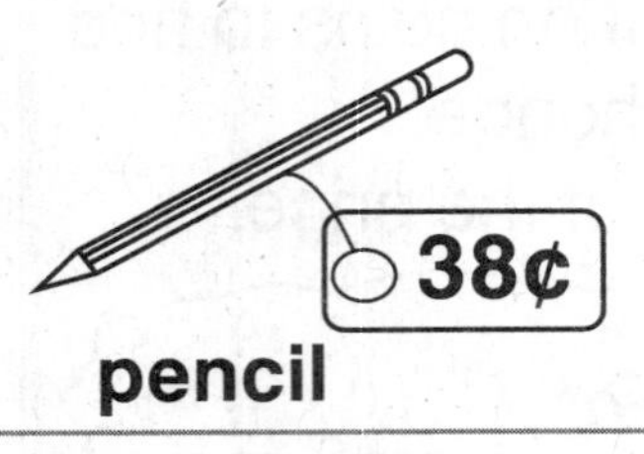

38¢

pencil

______, ______, ______

Your change is ______.

3. You have

____

You buy

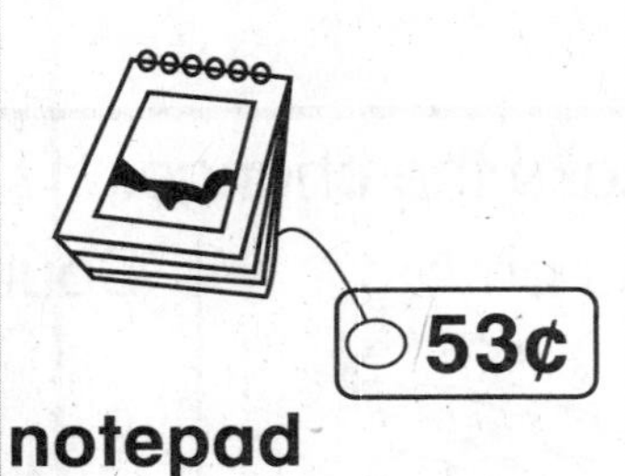

53¢

notepad

______, ______, ______

Your change is ______.

4. You have

____

You buy

ruler

70¢

______, ______, ______

Your change is ______.

Problem Solving

5. Meg buys a pencil for 60¢. She pays with 3 coins and gets 15¢ change. Draw the 3 coins Meg used to buy the pencil.

Name____

Identify Solid Figures

Some everyday objects are shaped like these solid **figures**.

cube

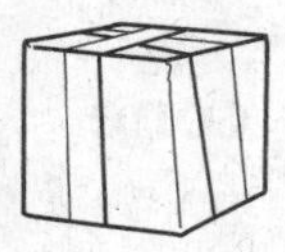

gift

block

cone

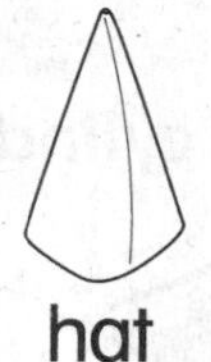

hat

funnel

cylinder

can

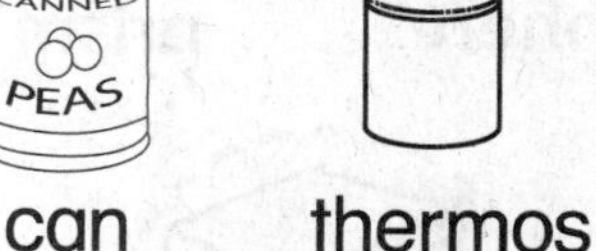

thermos

pyramid

tent pyramid

rectangular prism

cassette

sphere

baseball

marble

Circle the objects that are shaped like the solid figure. Cross out the objects that are not shaped like the solid figure.

1.

cone

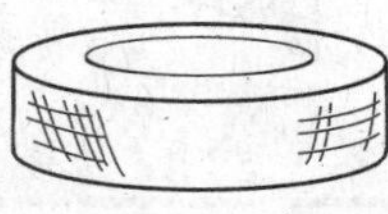

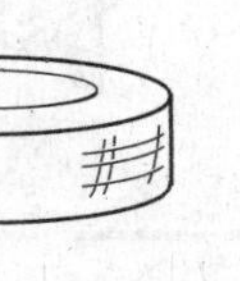

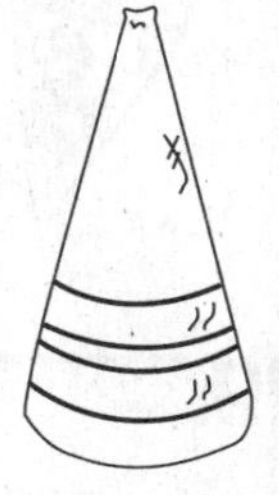

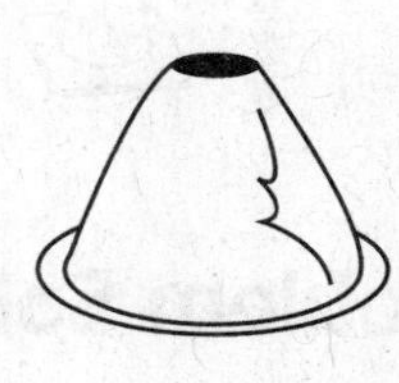

2.

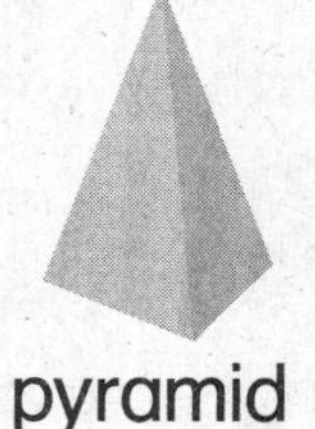

pyramid

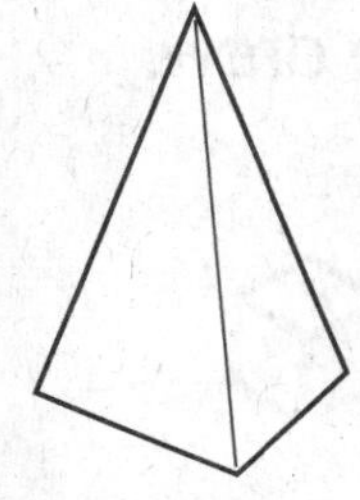

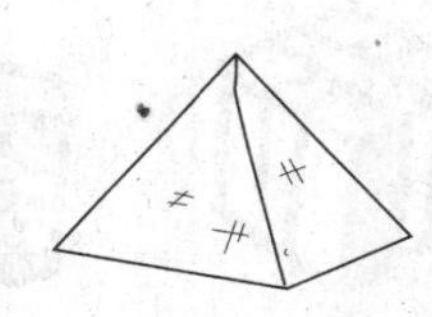

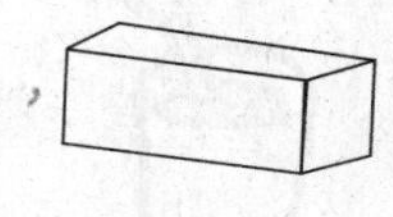

MG2.0 Students identity and describe the attributes of common figures in the plane and common objects in space: **MG2.1** Describe and classify plane and solid geometric shapes (e.g., circle, triangle, square, rectangle, sphere, pyramid, cube, rectangular prism) according to the number and shape of faces, edges, and vertices.

Name________________________________

Identify Solid Figures

Use the solid figures to help you remember.

Circle the objects that have the same shape.
Cross out the objects that do not belong.
Name the solid figures you circled.

sphere

rectangular prism

pyramid

cylinder

cube

cone

1.    

cube

2. 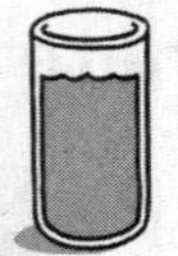  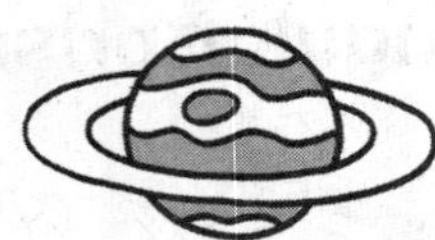 

____

3. 

____

4.  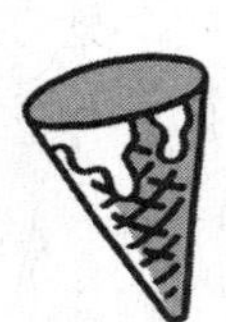  

____

Problem Solving

5. Jane drew a solid figure. It is not a cube.
It is not a cylinder. Circle the figure Jane drew.

Name______________________________

Attributes of Solid Figures

A **face** is a flat side of a solid figure.

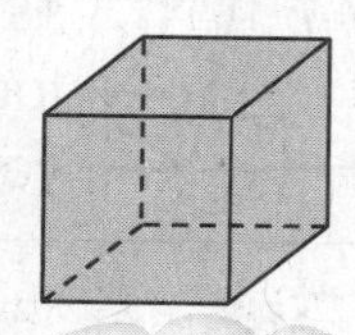

A cube has 6 faces.

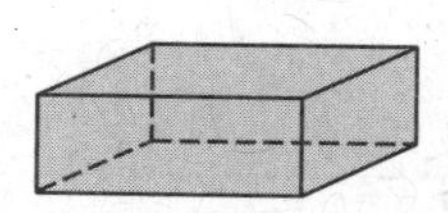

A rectangular prism has 6 faces.

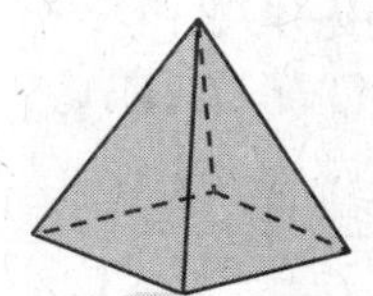

A pyramid has 5 faces.

An **edge** is formed where two faces meet.

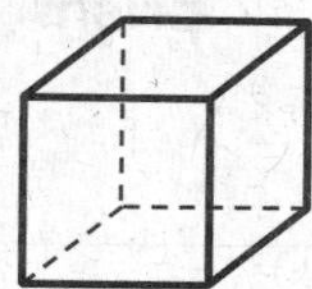

A cube has 12 edges.

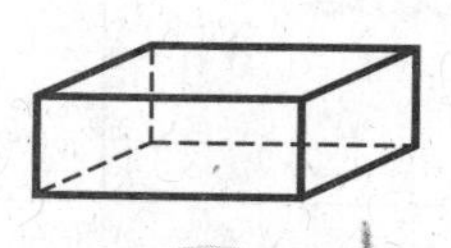

A rectangular prism has 12 edges.

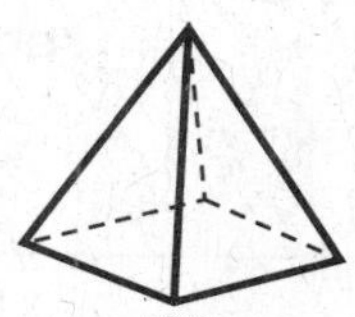

A pyramid has 8 edges.

A **vertex** is a corner where three or more edges meet. Two or more corners are called **vertices**.

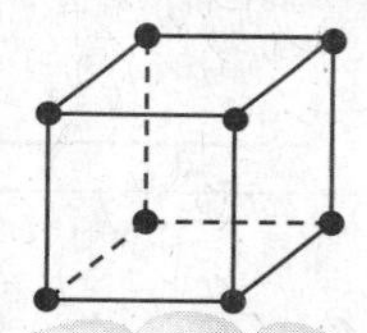

A cube has 8 vertices.

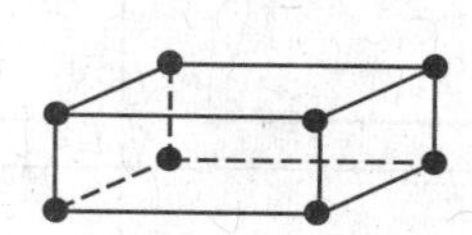

A rectangular prism has 8 vertices.

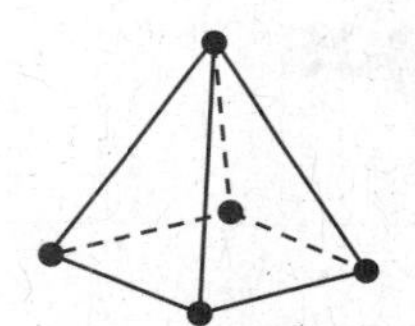

A pyramid has 5 vertices.

Use solid figures. Color the figure that matches the number of faces, edges, and vertices.

1. 6 faces, 12 edges, 8 vertices

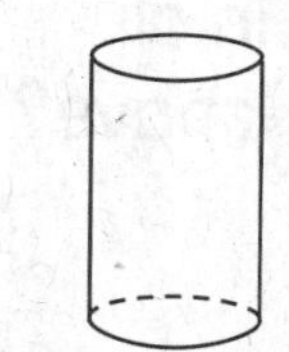

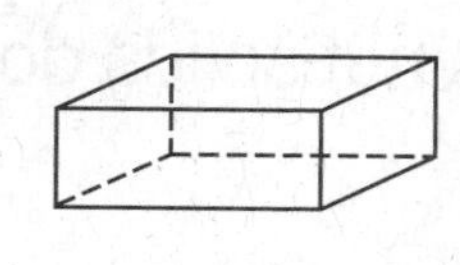

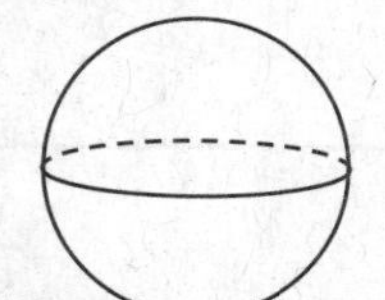

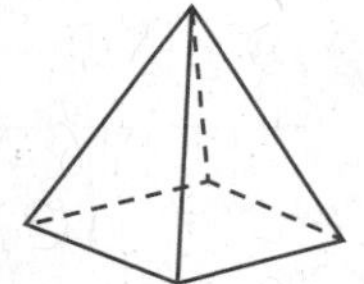

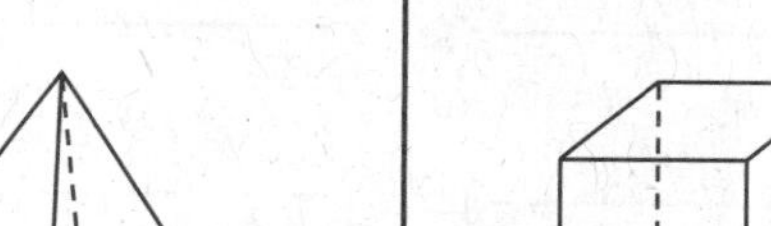

2. 5 faces, 8 edges, 5 vertices

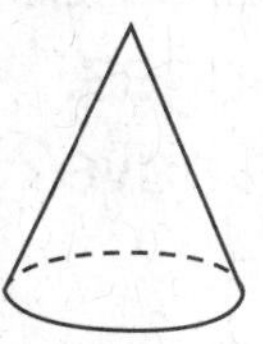

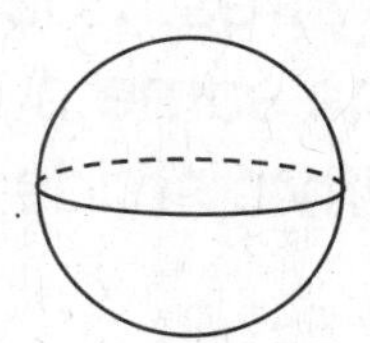

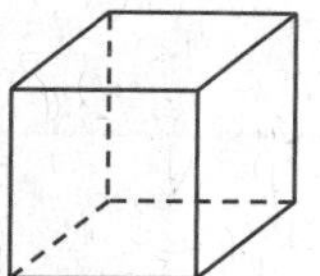

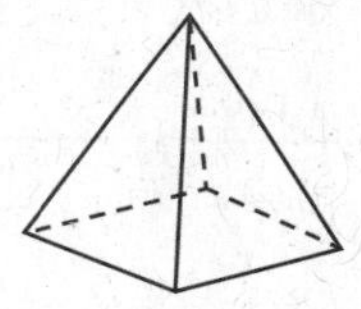

MG2.1 Describe and classify plane and solid geometric shapes (e.g., circle, triangle, square, rectangle, sphere, pyramid, cube, rectangular prism) according to the number and shapes of faces, edges, and vertices. **MG2.0** Students identify and describe the attributes of common figures in the plane and common objects in space:

Name________________________________

Lesson 12.3

Hands On: Attributes of Solid Figures

Use solid figures. Complete the table.
Write how many.

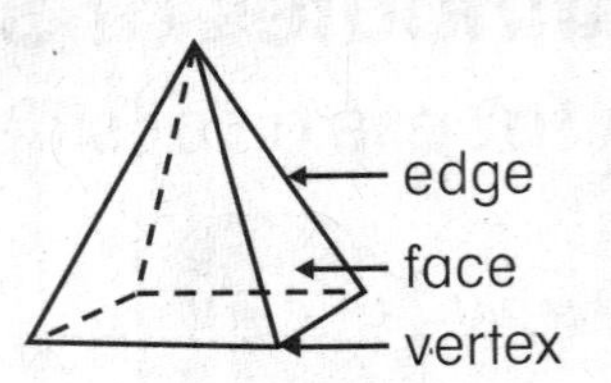

	1.	2.	3.	4.
solid figure	rectangular prism	cube	pyramid	rectangular prism
faces	6	______	______	______
edges	12	______	______	______
vertices	8	______	______	______

Problem Solving

5. Wen has two different solid figures. Each solid figure has the same number of edges. What solids does Wen have?

____

____

6. Clark has 2 solid figures. His figures have 10 faces in all. What solids does Clark have?

____

____

Make Plane Figures from Solid Figures

You can draw around the faces to find the plane figures.The faces of solid figures are plane figures.

A cube has 6 faces.
They are all squares.

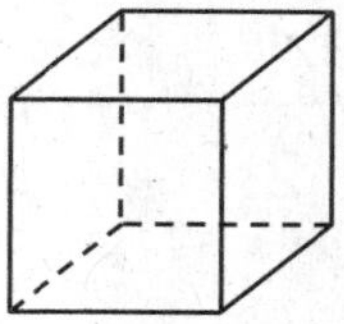

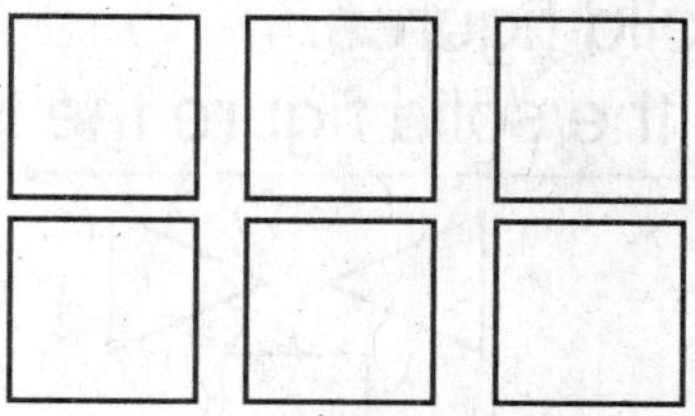

A rectangular prism has 6 faces. Its faces are all rectangles.

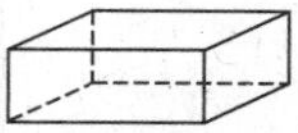

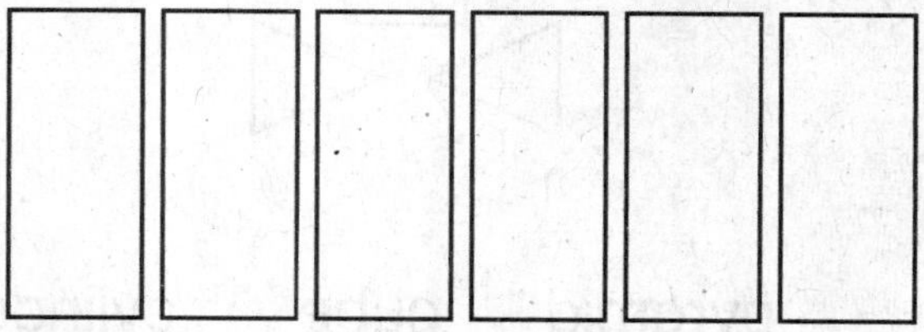

A pyramid has 5 faces.
One face is a square.
Four faces are triangles.

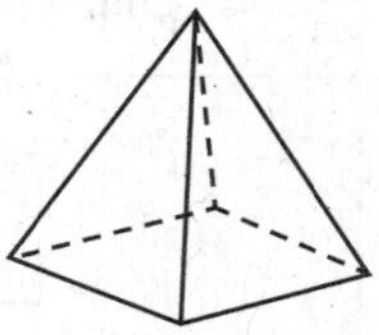

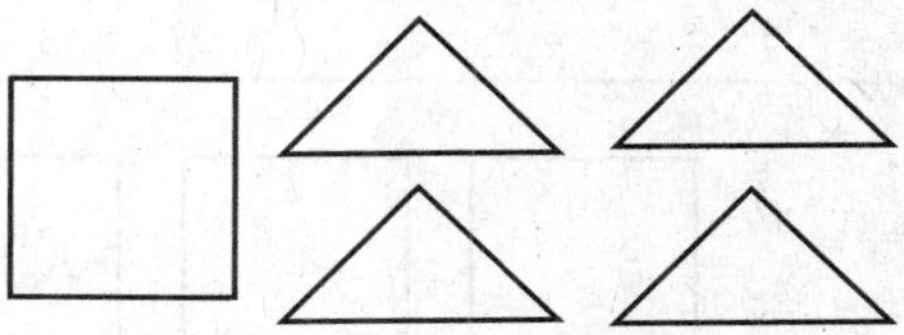

Use the solid figure. Draw around the faces.
Circle the faces of the solid figure.

1.

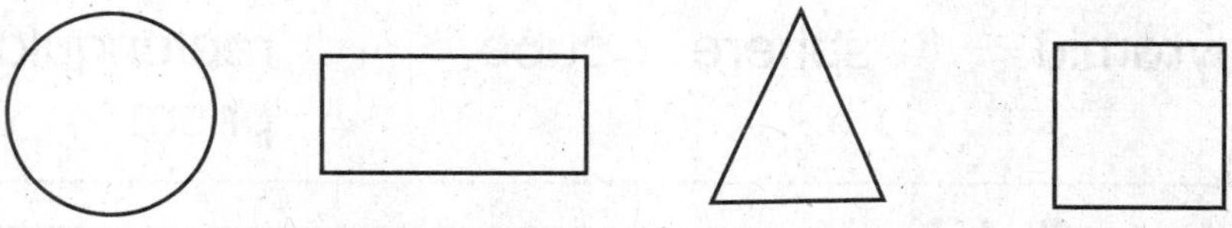

2.

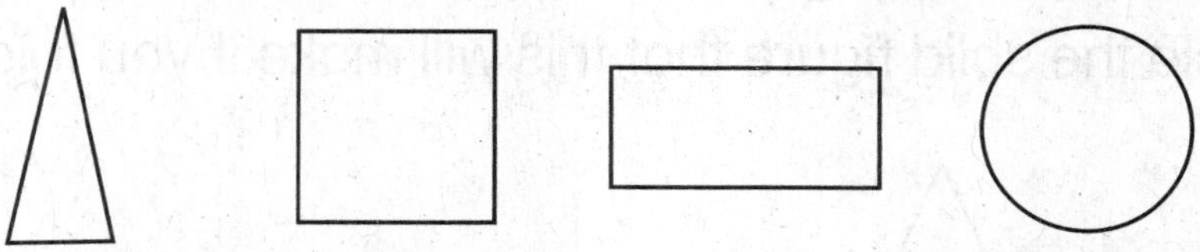

3.

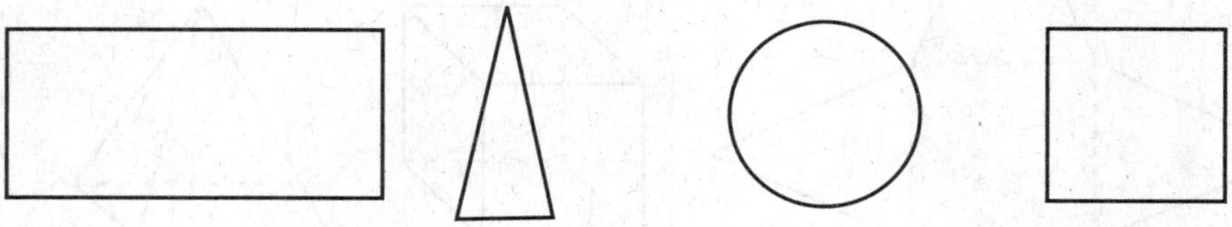

Name____

Hands On: Make Plane Shapes from Solid Figures

Use solid figures.
Circle the solid figure the faces make.

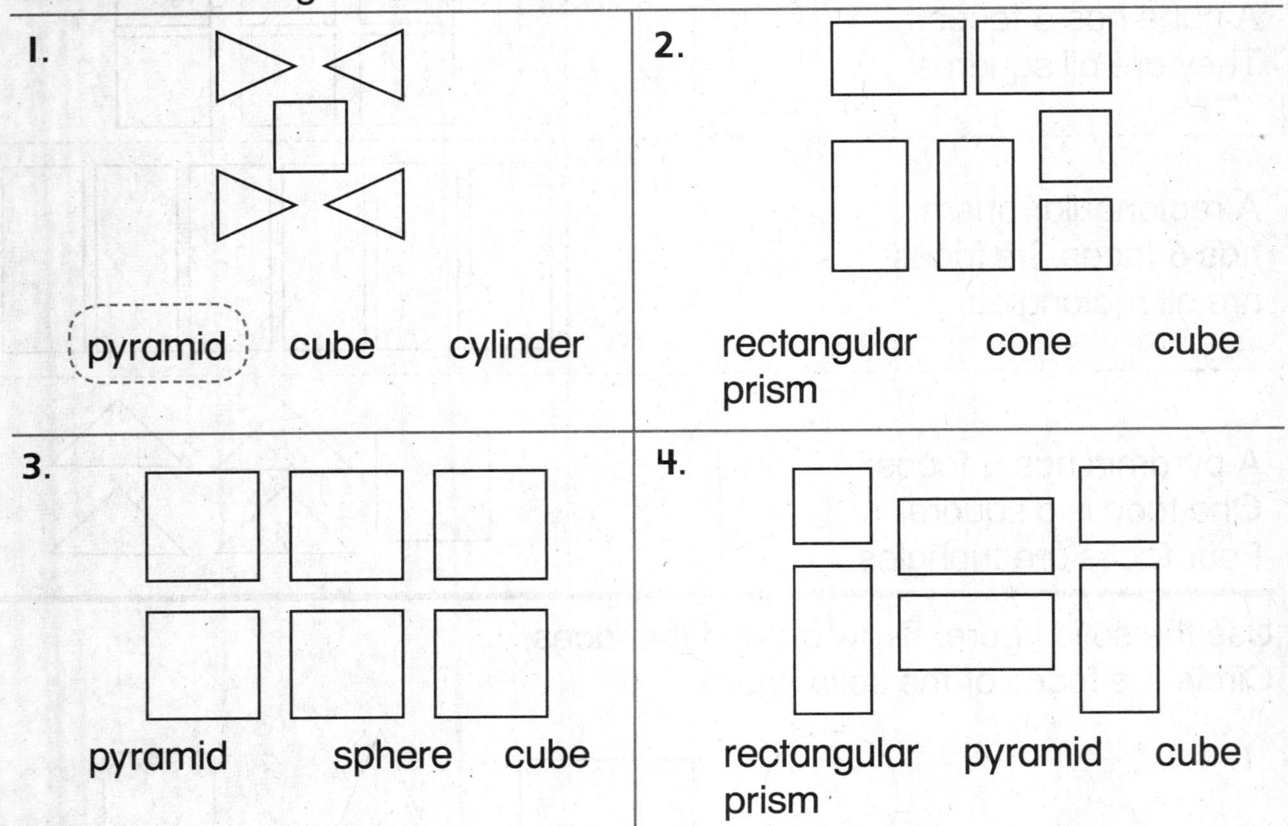

1. pyramid cube cylinder

2. rectangular prism cone cube

3. pyramid sphere cube

4. rectangular prism pyramid cube

Problem Solving

Circle the solid figure that this will make if you fold it and tape it together.

5.

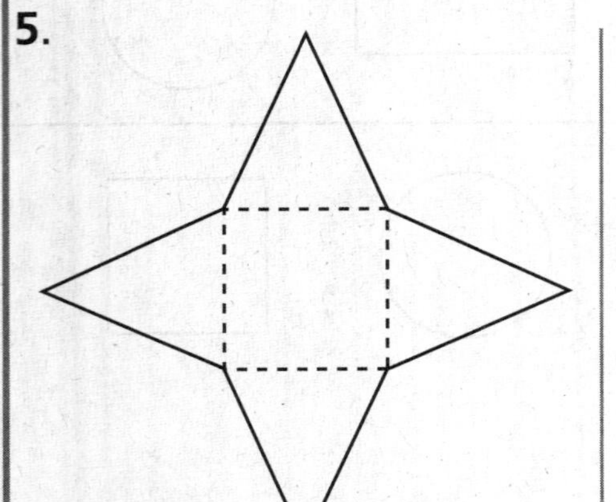

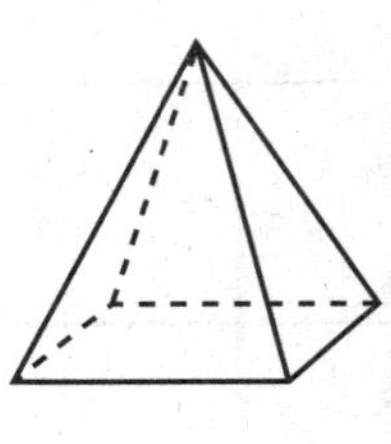

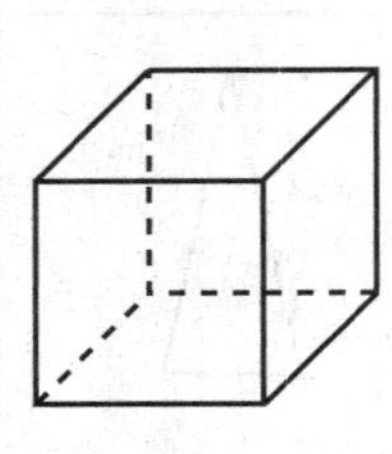

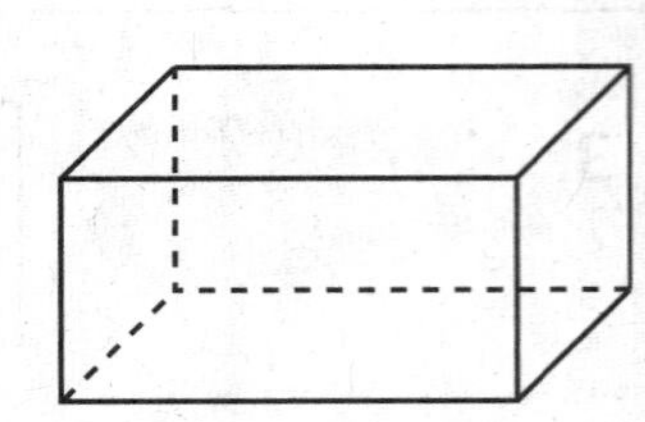

Name______________________________

Lesson 12.5

Problem Solving Strategy: Make a Table

Frank is making three pyramids.
How many triangles does he need
to make these three figures?

Read to Understand

1. **What do you want to find out?** Circle the important information. Underline the question.

Plan

2. **How will you solve the problem?** make a table

Solve

3. Complete the table to find the number of triangles.

Number of Triangles on Pyramids			
number of pyramids	1	2	3
number of triangles	4	8	12

Frank needs 12 triangles.

Check

4. Does your answer make sense?

Yes. I can count 12 triangles on 3 pyramids.

Complete the table to solve.

5. Jim is making five pyramids. How many squares does he need to make these five figures?

______________					
number of pyramids	1	2	3	4	5
______________					

Jim needs ______ squares.

MG2.0 Students identify and describe the attributes of common figures in the plane and of common objects in space: MG2.2 Put shapes together and take them apart to form other shapes (e.g., two congruent right triangles can be arranged to form a rectangle).

Name____

Strategy • Make a Table

Problem Solving Strategy Practice

Name and complete the table. Then add to solve.

1. Tasha is making three pyramids. How many triangles does she need to make these three figures?

______________________________			
(pyramid)			
(triangle)			

Tasha needs ______ triangles.

2. Andy is making three cubes. How many squares does he need to make these three figures?

______________________________			
(cube)			
(square)			

Andy needs ______ squares.

Mixed Strategy Practice

Choose a strategy to solve.

Choose a Strategy
- Make a Table
- Draw a Picture
- Make a List

3. Juanita wants to make five cylinders. How many circles will she need to make these five solid figures?

Juanita will need ______ circles.

Name ______________________________ **Week 21**

Spiral Review

Color one part red.
Write the fraction that names the red part.

1. 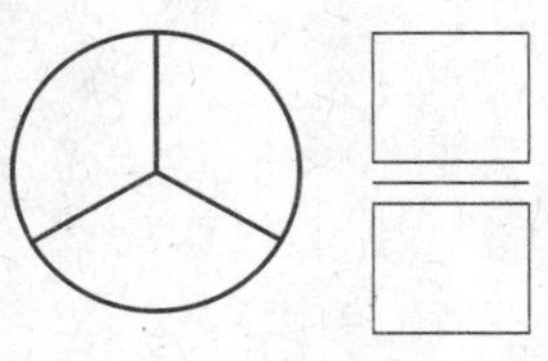

2. 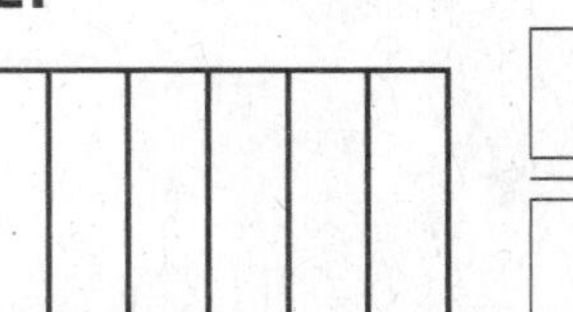

3. 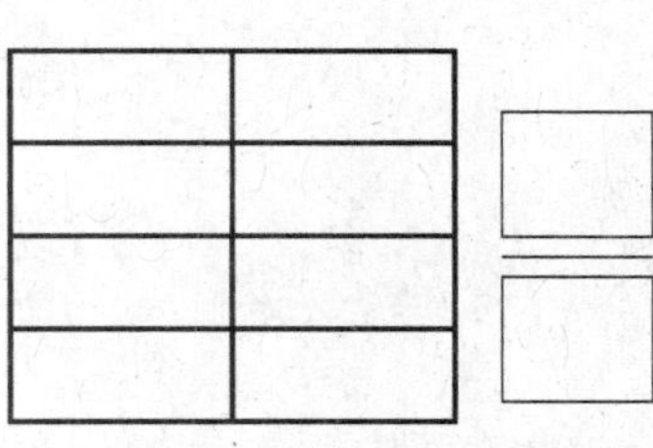

Use the picture graph.

Coins in Tom's Pocket					
pennies	●	●	●	●	●
nickels	●	●			
dimes	●	●	●		

Key: Each ● stands for 1.

4. How many more pennies than nickels are in Tom's pocket?

_____ more pennies

Write **yes** or **no** to tell about the surface of each object.

	Object	Does it slide?	Does it roll?	Does it slide and roll?
5.		_____	_____	_____

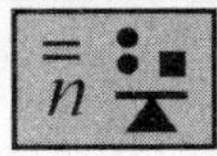

Use the table to solve.

Vegetables in the Garden	
Vegetable	**Number**
radishes	15
tomatoes	27
squash	24
carrots	31

6. How many more carrots than radishes are there?

_____ more carrots

Name____

Lesson 13.3

Combine Plane Figures

You can put figures together to make new figures.

Start with 2 squares.	Put them together.	Make a rectangle.

Use pattern blocks. Combine sides of figures to make a new figure. Trace the new figure.

1. 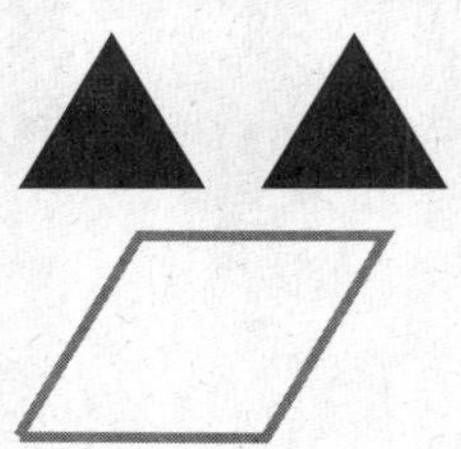

2. 

3. 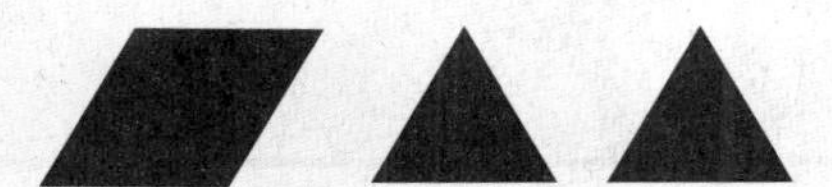

4. 

MG2.2 Put shapes together and take them apart to form other shapes (e.g., two congruent triangles can be arranged to form a rectangle). **MG2.0** Students identify and describe the attributes of common figures in the plane and of common objects in space:

Name______________________________________

Combine Plane Figures

Use pattern blocks. Combine sides of figures to make a new figure. Trace the new figure.

1. 

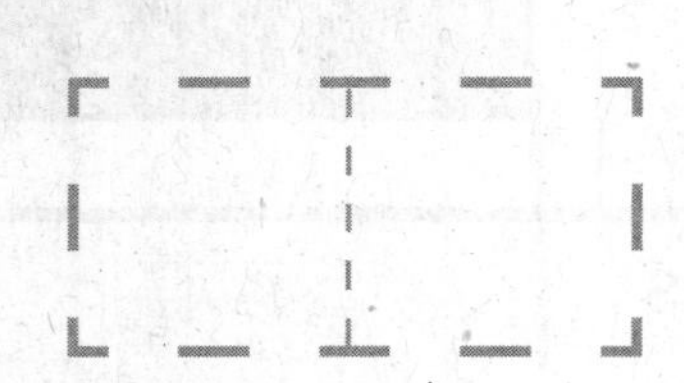

2. 

3.

4. 

Problem Solving

5. Circle a figure you can make using ▱, ▱, and △.

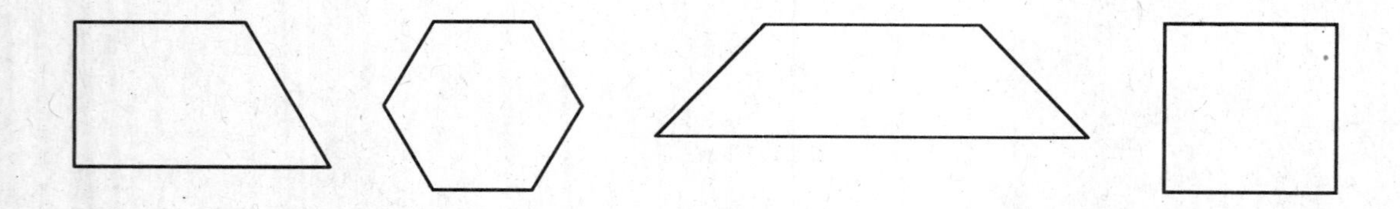

Name

Separate Plane Figures

You can take figures apart to make other figures.

Start with a hexagon.	Draw a line.
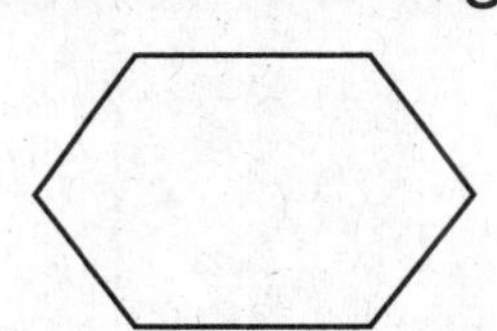	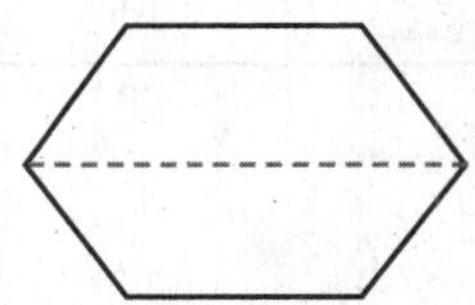
Fold along the line.	Cut the hexagon along the fold to make two figures.
	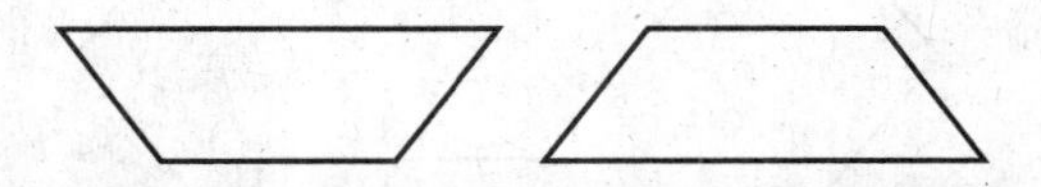

Use paper figures. Fold to make two new figures.
Cut along the fold. Trace the new figures.

1. 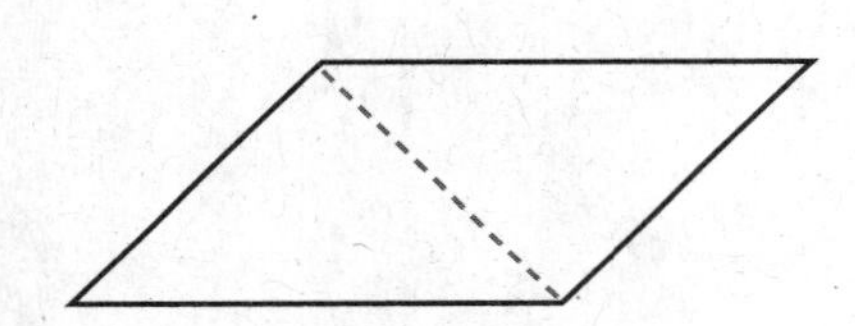

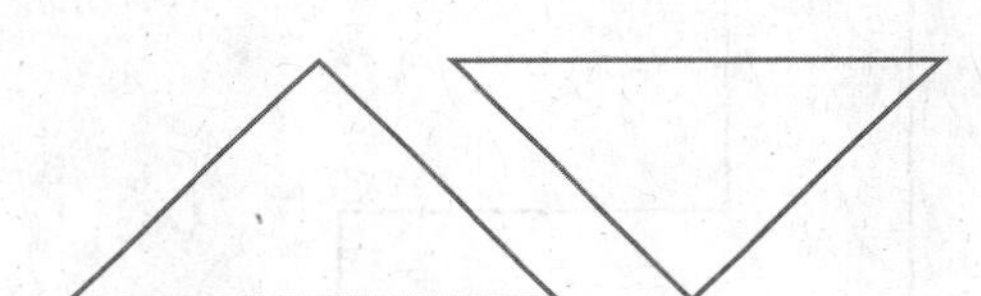

2. 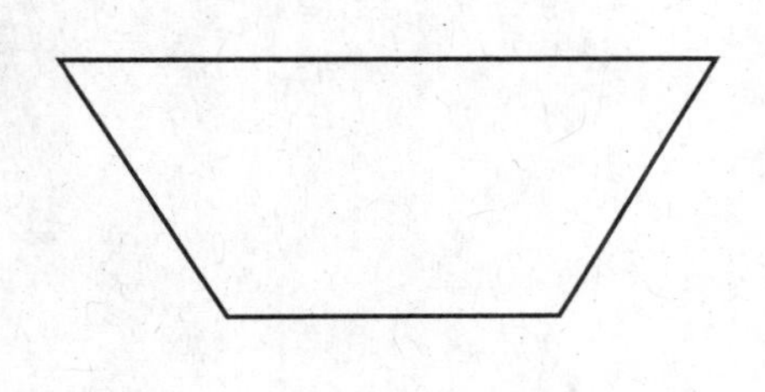

3. 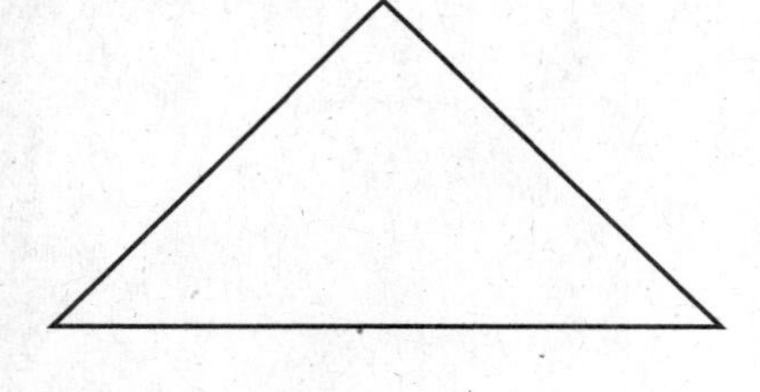

MG2.2 Put shapes together and take them apart to form other shapes (e.g., two congruent triangles can be arranged to form a rectangle). MG2.0 Students identify and describe attributes of common figures in the plane of common objects in space.

Name___________________________________ **Lesson 13.4**

Separate Plane Figures

Use paper figures. Fold to make three new figures. Cut along the folds. Trace the new figures.

1.

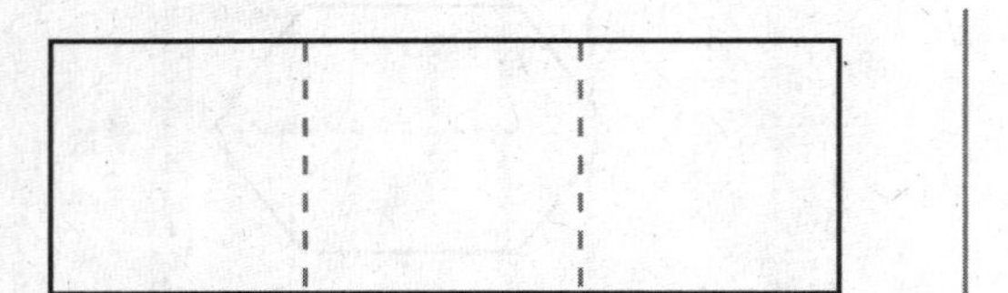

2.

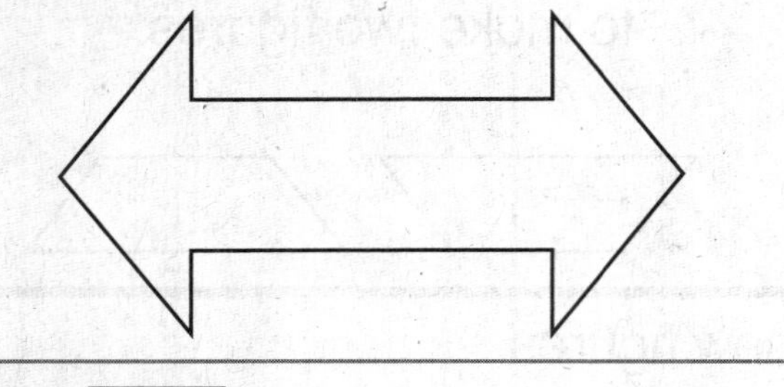

3.

4.

Problem Solving

5. Jeff wants to make a plane figure. He uses 2 colors of pattern blocks. Draw and color a figure that Jeff could make.

Name____________________

Lesson 13.5

Problem Solving Workshop Strategy: Use Logical Reasoning

I am a plane figure. I have fewer than 6 vertices. I have fewer than 4 sides. Which figure am I?

Read to Understand

1. Circle the information. Underline the question.

Plan

2. How do you solve the problem?

use logical reasoning

Solve

3. Use the information. Cross out the figures that do not fit the information. Circle the figure that answers the question.

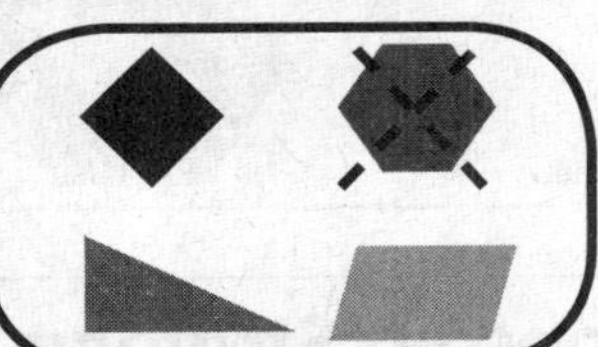

This figure does not have fewer than 6 vertices.

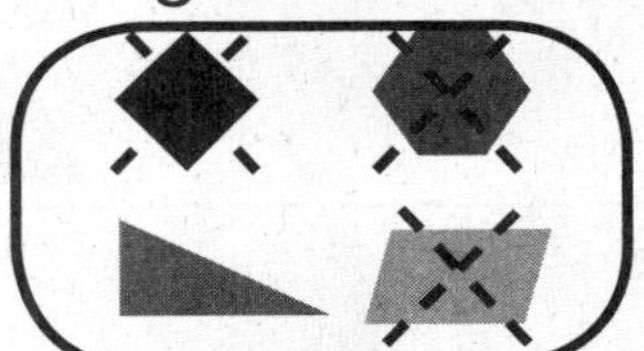

These figures do not have fewer than 4 sides.

The triangle answers the question.

Check

4. Does your answer make sense? Yes. The triangle fits all of the clues.

Use logical reasoning to solve.
Cross out the figures that do not fit the information.
Circle the figure that answers the question.

5. I am a plane figure with more than 3 vertices. I have fewer than 5 sides. Which figure am I?

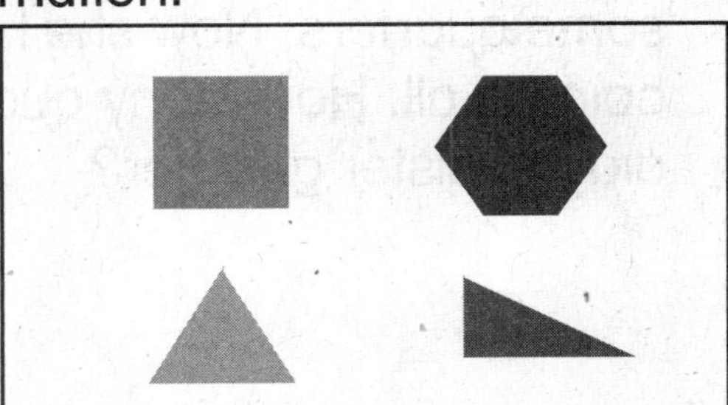

MG 2.1 Describe and classify plane and solid geometric shapes (e.g., circle, triangle, square, rectangle, sphere, pyramid, cube, rectangular prism) according to the number and shape of faces, edges, and vertices.

Name____ Lesson 13.5

Problem Solving Strategy: Use Logical Reasoning

Use logical reasoning to solve.
Cross out the figures that do not fit the information.
Circle the figure that answers the question.

1. I am a plane figure.
I have 3 vertices.
Which figure am I?

2. I am a plane figure.
I have more than 3 sides.
I have fewer than 5 vertices.
Which figure am I?

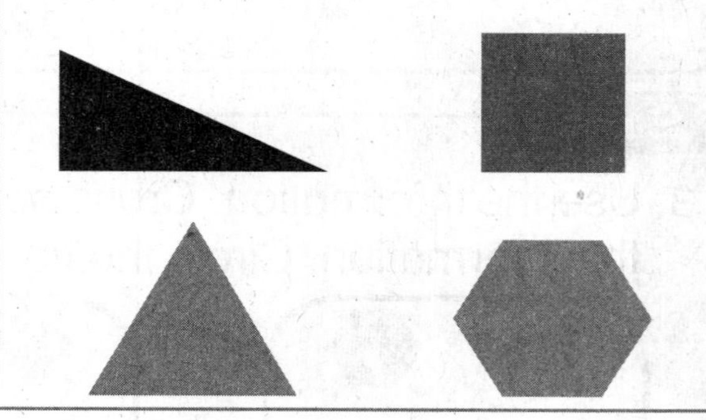

Mixed Strategy Practice

Choose a strategy to solve.

Choose a Strategy
- Use Logical Reasoning
- Draw a Picture
- Act it Out

3. Sheila has 2 quarters and 1 dime. Her sister gives her 3 nickels and a penny. How much money does Sheila have now?

__________¢

4. Doris has 3 pennies and 5 nickels. Her sister gives her some quarters. Now she has 12 coins in all. How many quarters did her sister give her?

__________ quarters

Spiral Review

Read the number. Write the number in different ways.

1. three hundred forty-nine

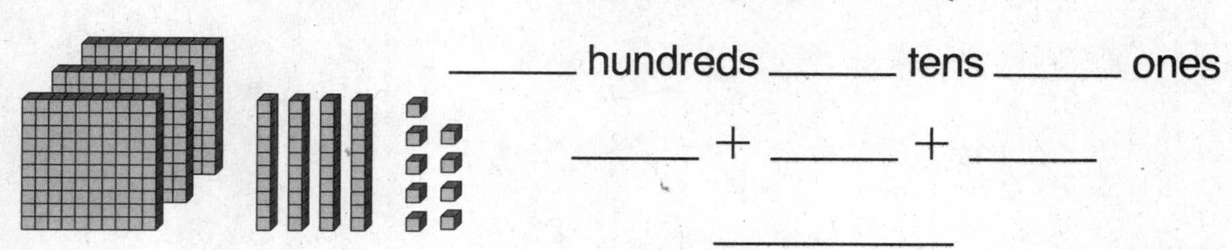

_____ hundreds _____ tens _____ ones

_____ + _____ + _____

_____

Write a title and labels for the bar graph then complete the bar graph to show the information.

2. Ali took a survey of her classmates' favorite fruits. 4 classmates chose apple, 5 chose peach, 6 chose pear, and 2 chose plum.

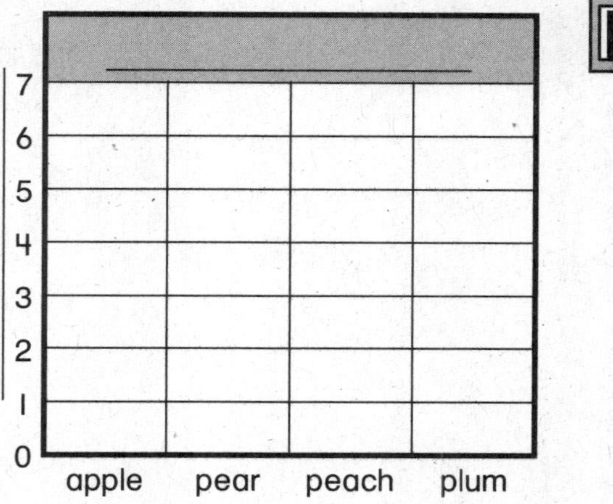

Circle the objects that have the same figure. Cross out the objects that do not belong. Name the solid figures you circled.

3.   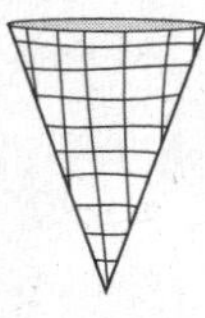 

_____

Draw a line through the information you do not need. Then solve.

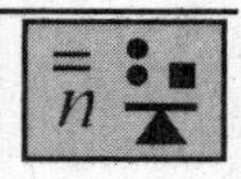

4. On Friday, 156 people visit the art museum. On Saturday, 117 people visit the science museum and 263 people visit the art museum. How many people visit a museum on Saturday?

_____ people

Spiral Review

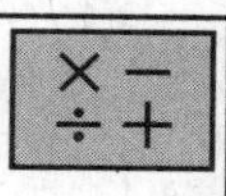

Use Workmat 6 and ●. Draw to show your work. Write how many in each group.

1. Divide 12 ● into 4 equal groups.

_____ in each group

Use cubes to show the number as tens and ones. Draw what you built. Write **even** or **odd**.

2. 12

____

Measure the length to the nearest inch.

3.

Measure: about _____ inches

4.

Measure: about _____ inches

Use Workmat 5 and 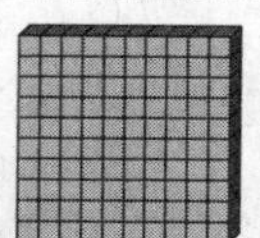 ▭ ▫. Subtract.

5.

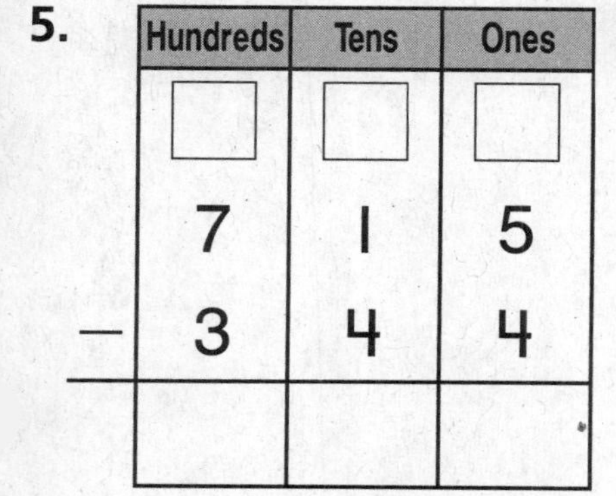

	Hundreds	Tens	Ones
	☐	☐	☐
	7	1	5
−	3	4	4

6.

	Hundreds	Tens	Ones
	☐	☐	☐
	5	2	7
−	1	6	3

7.

	Hundreds	Tens	Ones
	☐	☐	☐
	3	5	2
−	2	7	0

Name ______________________________ **Week 24**

Spiral Review

Write the product.

1.	2.	3.	4.	5.	6.
$6 \times 2 = $ ___	$10 \times 3 = $ ___	$5 \times 7 = $ ___	$10 \times 2 = $ ___	$3 \times 5 = $ ___	$2 \times 8 = $ ___

Use the line plot to answer the questions.

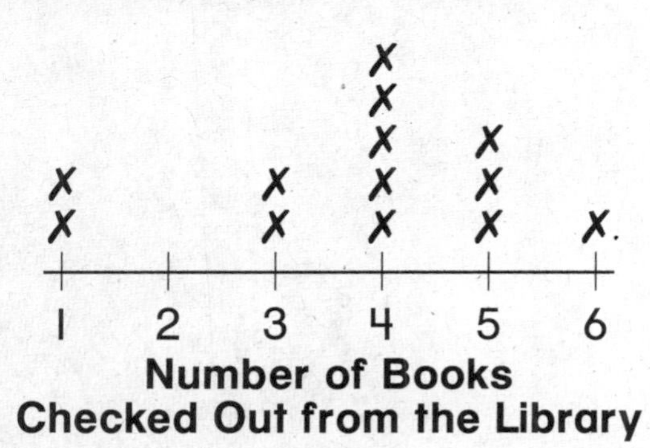

Number of Books Checked Out from the Library

7. Write a number sentence to find the range of the data.

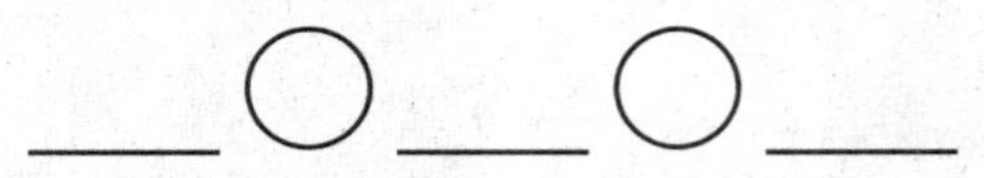

8. What is the mode of the data?

The mode of the data is _____.

Use cubes. Measure the real object.

9.

Find the object.	Measure.
	about _____ cubes

Find a rule. Complete the table.

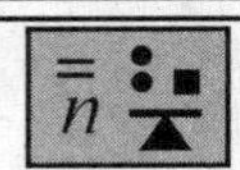

10. **Rule: Add _____.**

In	Out
3	
4	11
5	
6	13

11. **Rule: Add _____.**

In	Out
2	7
4	9
6	
8	

Name

Lesson 15.4

Problem Solving Strategy: Make a Model

Ed and Jan each have 5 pens. $\frac{3}{5}$ of Ed's pens are blue. $\frac{5}{5}$ of Jan's pens are blue. Who has more blue pens?

Read to Understand

1. What do you need to find?
Who has more blue pens, Ed or Jan?

Plan

2. What strategy can you use to solve the problem?
Make a model to compare the fractions.

Solve

3. Use squares to stand for the pens. Color to show how many blue pens each child has. Then compare.

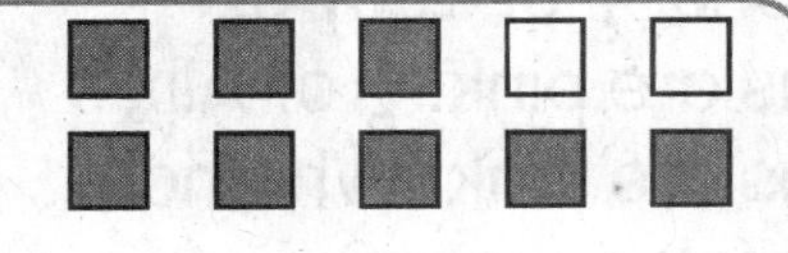

$\frac{3}{5}$ is less than $\frac{5}{5}$

Jan has more blue pens.

Check

4. Does your answer make sense?
Yes. My model shows that Jan has more blue pens.

Shade the squares to make a model.
Write is **greater than** or is **less than**.

5. Jim and Kat each have 6 apples. $\frac{3}{6}$ of Kat's apples are red. $\frac{4}{6}$ of Jim's are red. Who has more red apples?

□ □ □ □ □ □
□ □ □ □ □ □

$\frac{3}{6}$ ________ $\frac{4}{6}$.

________ has more red apples.

NS 4.2 Recognize fractions of a whole and parts of a group (e.g., one–fourth of a pie, two–thirds of 15 balls). **NS 4.0** Students understand that fractions and decimals may refer to parts of a set and parts of a whole. **MR 1.2** Use tools, such as manipulatives or sketches, to model problems.

Name ________________________________

Lesson 15.4

Problem Solving Strategy: Make a Model

Shade the squares to make a model.
Write **is greater than** or **is less than**.

1. Greg and Jeff each have 5 hats. $\frac{2}{5}$ of Greg's hats are gray. $\frac{4}{5}$ of Jeff's hats are gray. Who has more gray hats?

   $\frac{2}{5}$ is less than $\frac{4}{5}$.

   Jeff has more gray hats.

2. Lucy and Alix each have 6 ribbons. $\frac{3}{6}$ of Lucy's ribbons are pink. $\frac{2}{6}$ of Alix's ribbons are pink. Who has more pink ribbons?

   $\frac{3}{6}$ ____________ $\frac{2}{6}$.

   ________ has more pink ribbons.

3. Cam and Bo each have 8 crayons. $\frac{5}{8}$ of Cam's crayons are red. $\frac{6}{8}$ of Bo's crayons are red. Who has more red crayons?

   $\frac{5}{8}$ ____________ $\frac{6}{8}$.

   ________ has more red crayons.

4. Roy and Liz each have 5 grapes. $\frac{4}{5}$ of Roy's grapes are green. $\frac{3}{5}$ of Liz's grapes are green. Who has more green grapes?

   $\frac{4}{5}$ ____________ $\frac{3}{5}$.

   ________ has more green grapes.

Name ______________________________ **Week 25**

Spiral Review

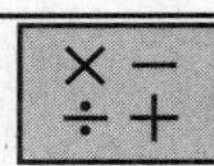

Draw equal groups. Skip-count to find how many in all.
Write the multiplication sentence.

1. 5 equal groups. 3 ● in each group.

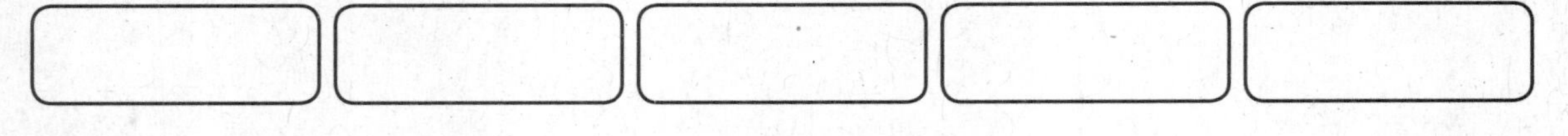

____, ____, ____, ____, ____ ____ × ____ = ____

Use the bar graph to answer the questions.

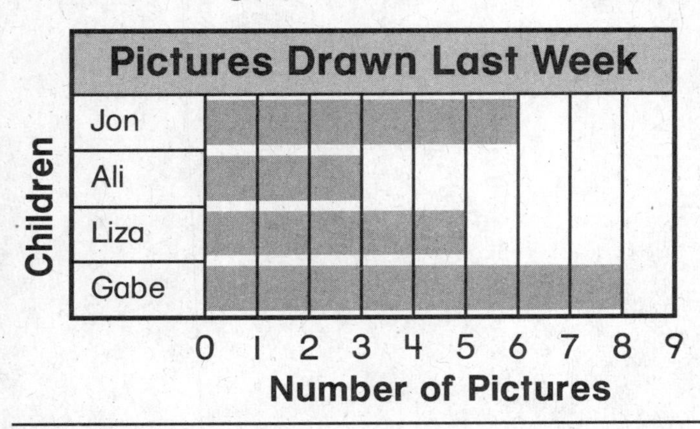

2. How may pictures did Ali and Gabe draw last week together?

____ pictures

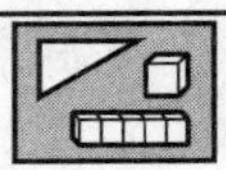

Measure the length to the nearest inch.

3.

Measure: about ____ inches

4.

Measure: about ____ inches

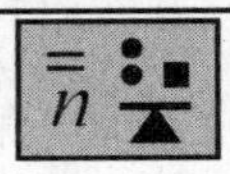

Draw a picture to solve. Write a number sentence.

5. Arlo has 4 bags of pretzels. He is given 6 more bags. How many bags does he have in all?

____ bags

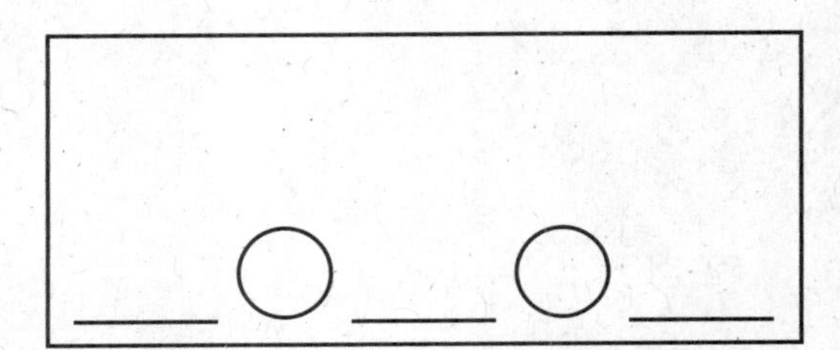

Name______________________________

Lesson 16.2

Hundreds, Tens, and Ones

How many in all?

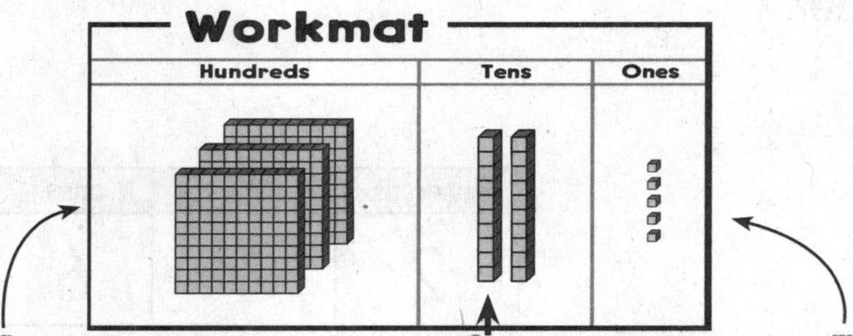

There are 3 hundreds. There are 2 tens. There are 5 ones.

You can use a chart to write how many.

hundreds	tens	ones
3	2	5

3 hundreds 2 tens 5 ones is the same as 325.

Use Workmat 5 and .
Write how many hundreds, tens, and ones. Write the number.

1. 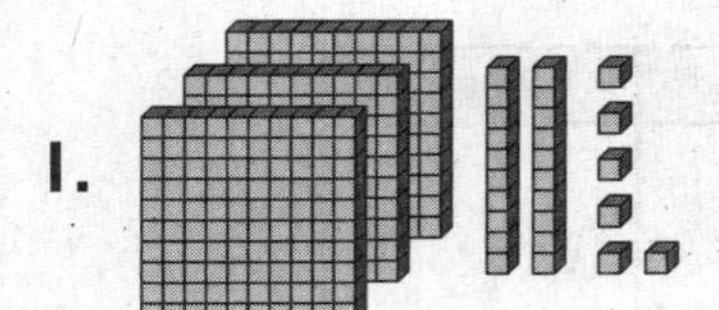

hundreds	tens	ones

______

Think: the digit 0 in the ones place means that there are 0 ones.

2.

hundreds	tens	ones

______

3.

hundreds	tens	ones

______

NS1.2 — Use words, models, and expanded forms (e.g., 45 = 4 tens + 5) to represent numbers (to 1,000). NS1.0 — Students understand the relationship between numbers, quantities, and place value in whole number up to 1,000. NS1.1 — Count, read, and write whole numbers to 1,000 and identify the place value for each digit.

Name ____________________

Lesson 16.2

Hundreds, Tens, and Ones

Write how many hundreds, tens, and ones.
Then write the number.

1.

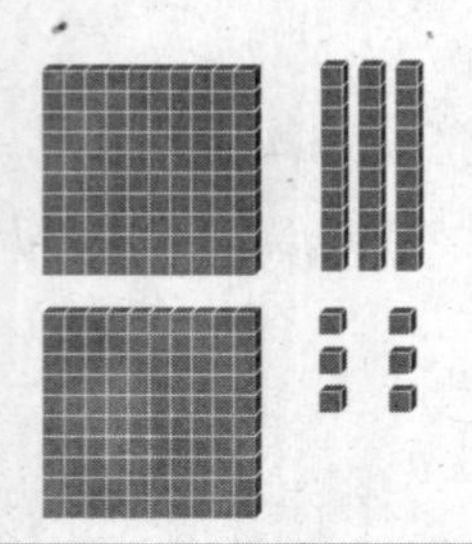

hundreds	tens	ones
2	3	6

236

2.

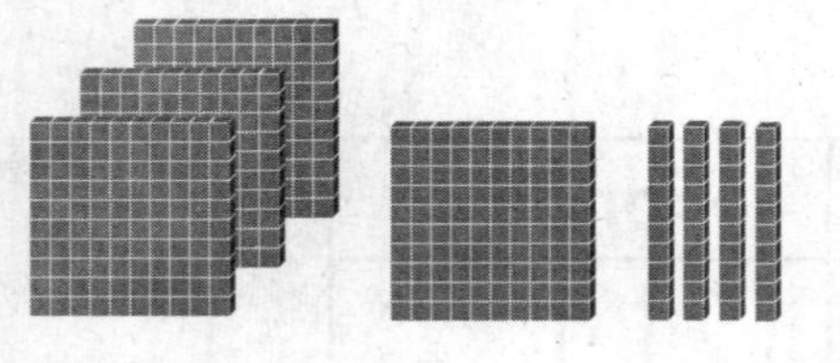

hundreds	tens	ones

______

3.

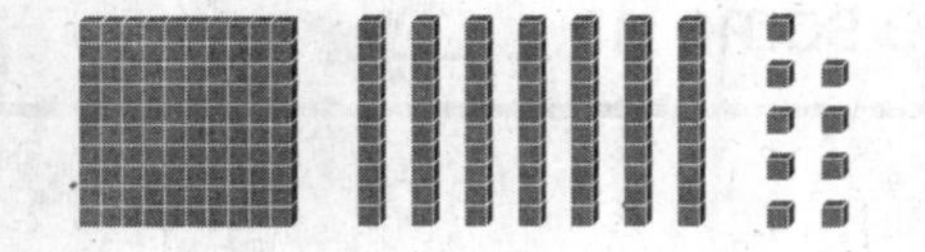

hundreds	tens	ones

______

4.

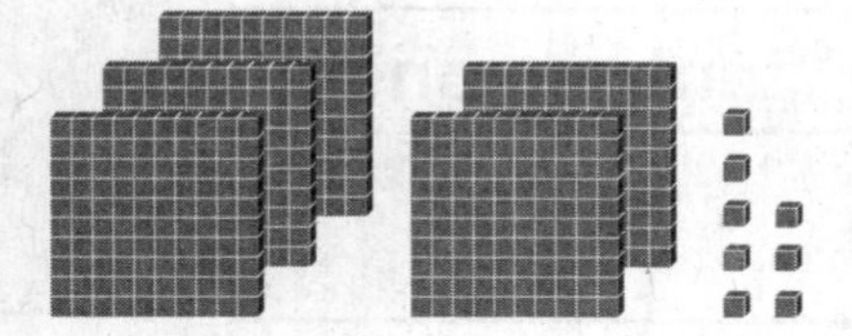

hundreds	tens	ones

______

Problem Solving

5. I have 8 ones, 3 tens, and 1 hundred. What number am I?

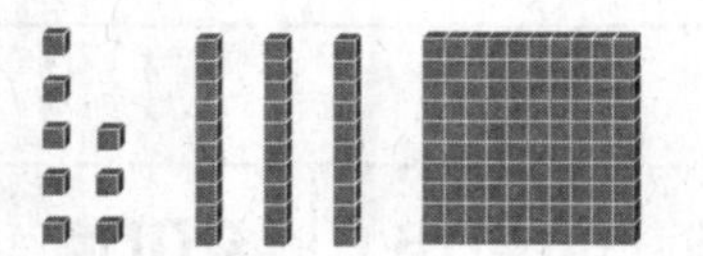

______

6. I have 6 ones, 1 ten, and 4 hundreds. What number am I?

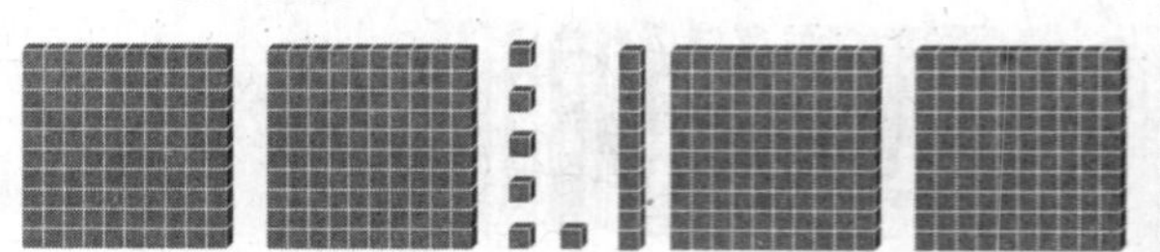

______

Name____ Lesson 16.5

Different Ways to Show Numbers

227 = 2 hundreds 2 tens 7 ones

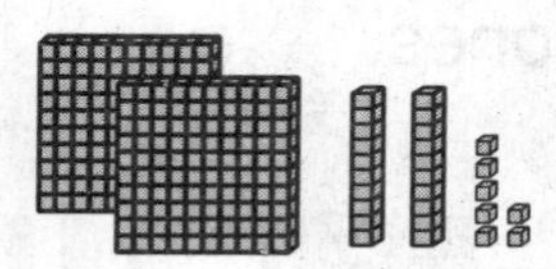

REMEMBER:
10 ones = 1 ten
10 tens = 1 hundred

You can show 227 in different ways.

This way trades 1 hundred for 10 tens.

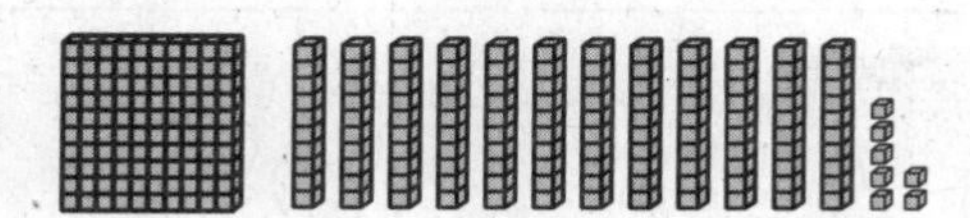

1 hundred 12 tens 7 ones

This way trades 1 ten for 10 ones.

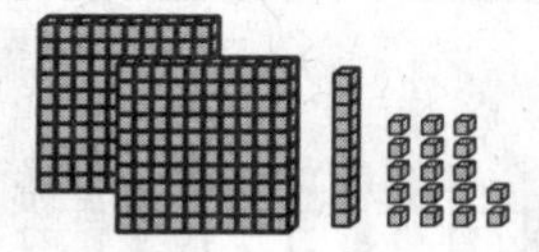

2 hundreds 1 ten 17 ones

Use [blocks].
Write how many hundreds, tens, and ones.

1. 461

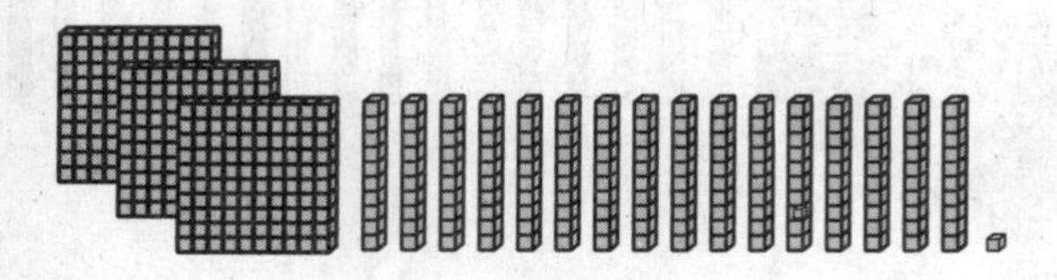

___ hundreds ___ tens ___ ones

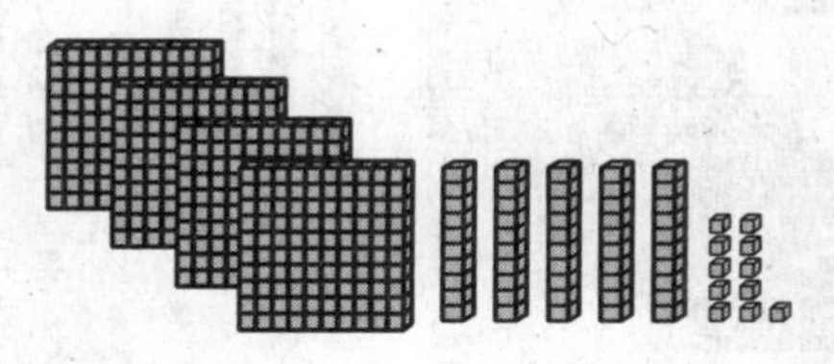

___ hundreds ___ tens ___ ones

2. 339

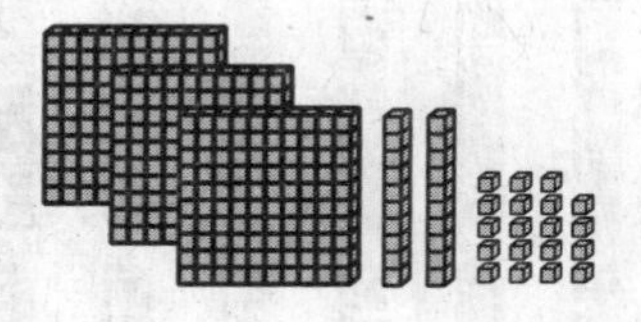

___ hundreds ___ tens ___ ones

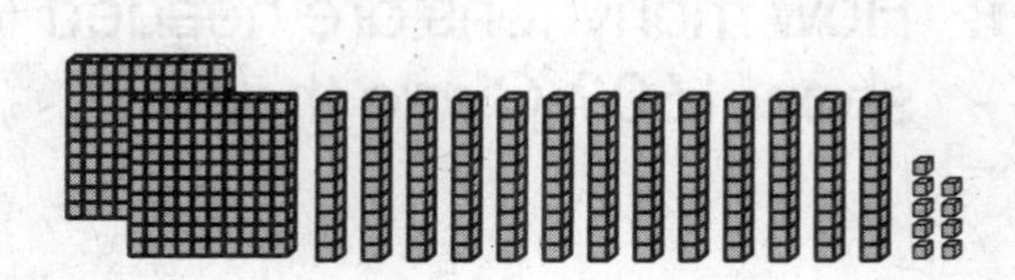

___ hundreds ___ tens ___ ones

NS1.2 Use words, models, and expanded forms (e.g., 45 = 4 tens + 5) to represent numbers (to 1,000). **NS1.0** Students understand the relationship between numbers, quantities, and place value in whole numbers up to 1,000. **MR3.0** Students note connections between one problem and another.

Name____

Lesson 16.5

Different Ways to Show Numbers

Use [hundred block] [ten rod] [one cube].
Write how many hundreds, tens, and ones.

1. 325

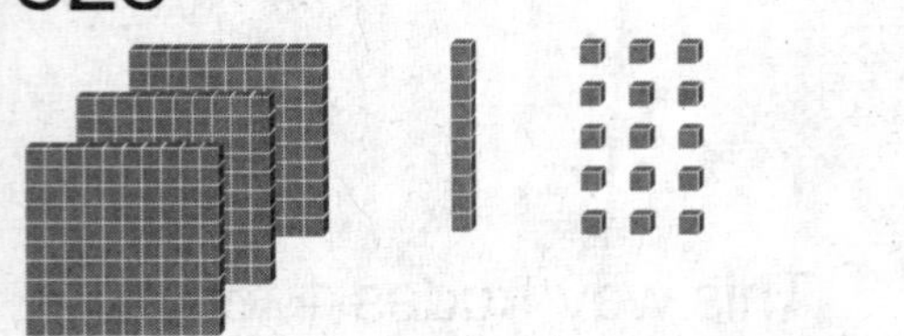

3 hundreds 1 tens 15 ones

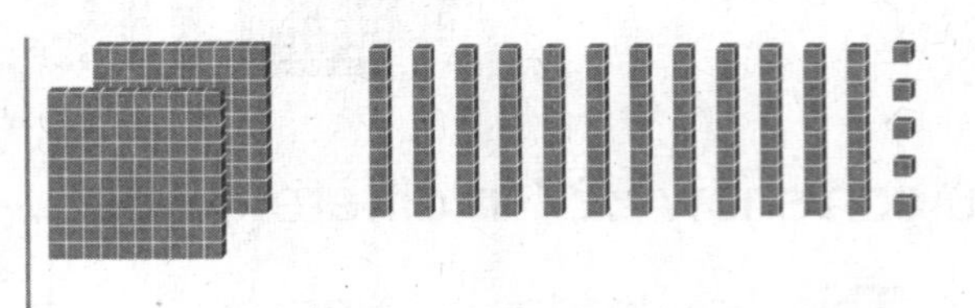

____ hundreds ____ tens ____ ones

2. 413

____ hundreds ____ tens ____ ones

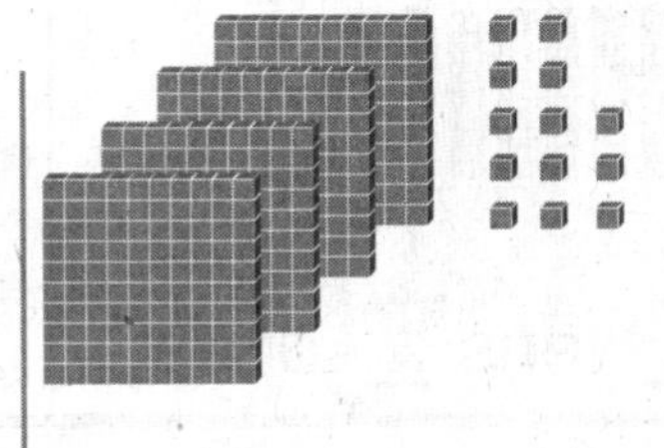

____ hundreds ____ tens ____ ones

3. 562

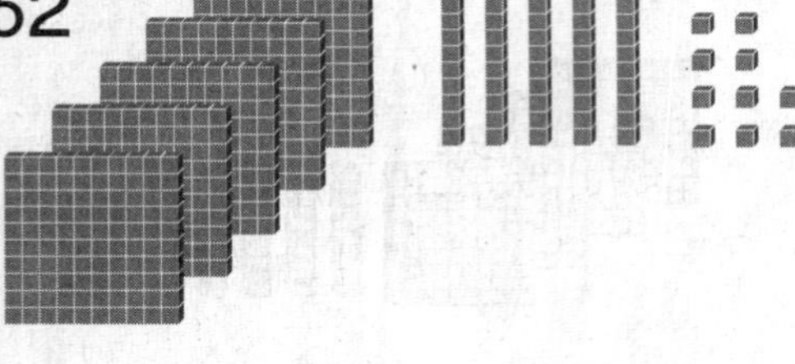

____ hundreds ____ tens ____ ones

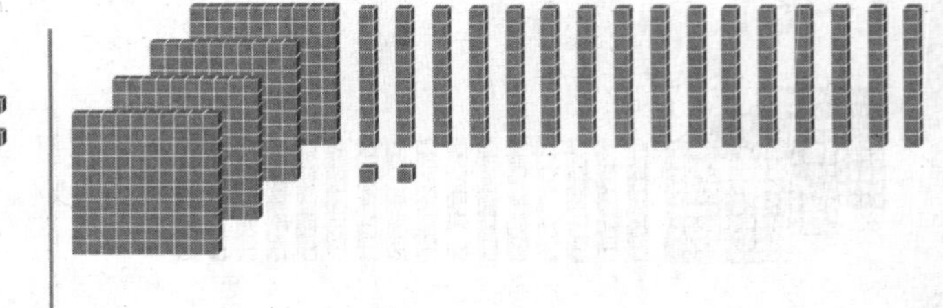

____ hundreds ____ tens ____ ones

Problem Solving

4. How many tens are needed to show 160? Circle them.

________ tens

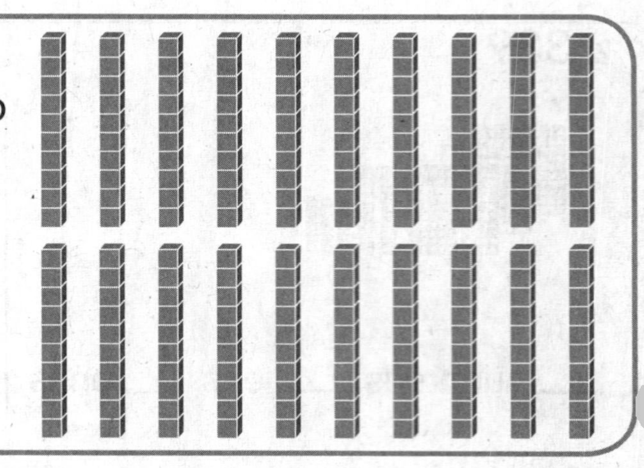

Name ______________________________ **Week 26**

Spiral Review

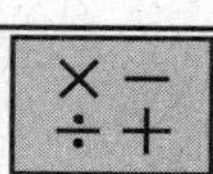

What fraction of the group are black?
What fraction of the group are gray?

1. 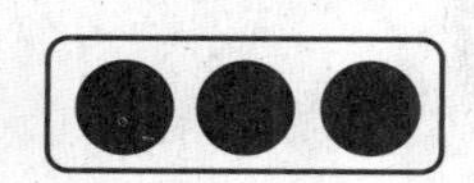 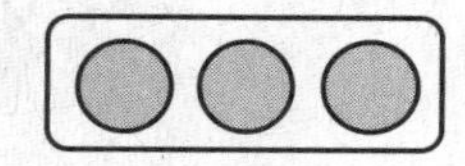 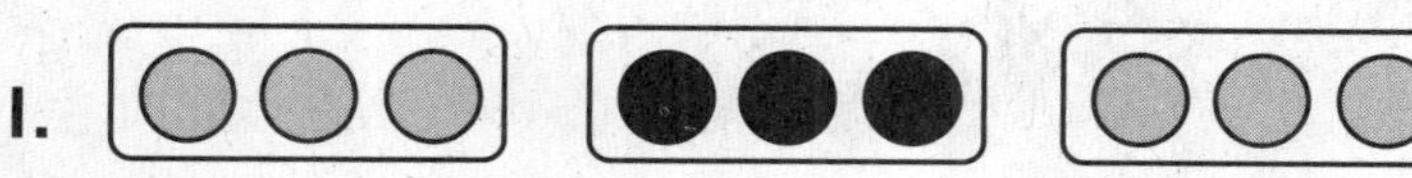

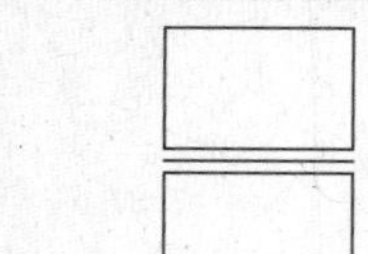 of the group are black 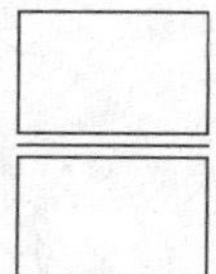 of the group are gray

Skip-count. Write the missing numbers.
Write a rule for the pattern.

2. 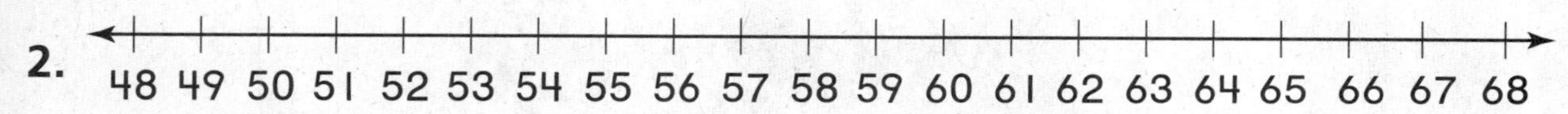

48 49 50 51 52 53 54 55 56 57 58 59 60 61 62 63 64 65 66 67 68

48, 51, 54, _____, _____ Rule: Count by __________.

Write the time in two ways.

3. 

_____ minutes after _____ o'clock

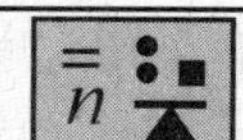

Use Workmat 11 and . Regroup if you need to. Write how many tens and ones. Write the difference.

4. Subtract 7 from 35.

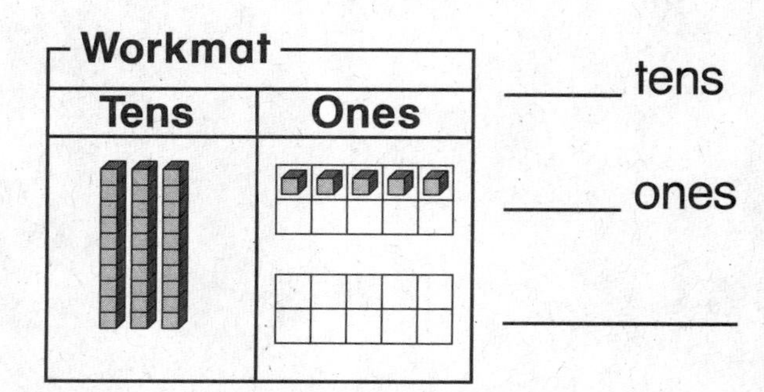

_____ tens

_____ ones

____

Name ___________________________

Lesson 16.7

Count Bills and Coins

Marya has 2 dollar bills, 1 quarter, 1 nickel, and 2 pennies. How much money does she have?

Remember: use a dollar sign ($) and a decimal point (.) to show money amounts.

Start with the bills. Then count on with the coins.

2 dollar bills		1 quarter	1 nickel	2 pennies	

Start with $1.00	Add $1.00	Then add 25¢	Count on 5¢	Count on 1¢	Count on 1¢

Say: $1.00, $2.00, $2.25, $2.30, $2.31 $2.32.

So Marya has a total of $2.32 .

Count on. Write the amount.

1. Paavo has 1 dollar bill, 4 dimes, and 1 nickel. How much money does he have?

____

2. Serena has 2 quarters, 1 dime, and 3 pennies. How much money does she have?

Hint: If there are no dollars, use zero as a place holder. $0.75 = 75¢

____

Name ______________________________ **Lesson 16.7**

Count Bills and Coins

Count on. Write the amount.

REMEMBER: Use a dollar sign ($) and a decimal point (.) to write money amounts.

1. Julio has 3 quarters, 1 nickel, and 1 penny. How much money does he have?

HINT: If there are no dollars, write 0.

$0.81

2. Van has 3 dollar bills, 3 dimes, and 2 pennies. How much money does he have?

____

3. Cassandra has 2 dollar bills, 2 quarters, and 1 dime. How much money does she have?

____

Problem Solving

4. Jim made these price tags. Cross out the prices that have the $ or the ¢ in the wrong place.

2.97$ $1.82

¢63 30¢

Name________________________________

Lesson 16.8

Problem Solving Strategy: Make a Model

Ryan has 1 dollar bill, 4 quarters, 1 dime, and 5 nickels. He wants to buy a keychain that costs $1.20. What are two different ways that he can pay the exact amount?

Read to Understand

1. What do you need to find out?

Plan

2. What strategy can you use? make a model

Solve

3. Make a model to show Ryan's money. Then show two different ways to make the amount.

Use 1 dollar bill, 1 dime, and 2 nickels to make $1.20.

Use 4 quarters, 1 dime, and 2 nickels to make $1.20.

Check

4. Does your answer make sense?
Yes. Both pictures show $1.20.

Use bills and coins to make a model.

5. Ivan has 1 dollar bill, 3 quarters, 2 dimes, and 1 nickel. He wants to buy a snack that costs $1.50. Draw two ways that he can pay the exact amount.

NS5.1 – Solve problems using combinations of coins and bills. **NS5.2** – Know and use the decimal notation and the dollar and cent symbols for money. **MR2.1** – Defend the reasoning used and justify the procedures selected.

Name______________________________

Strategy • Make a Model

Use bills and coins to make a model.

1. Raji has 2 dollar bills, 1 quarter, 3 dimes, and 1 nickel. He wants to buy a ball that costs $1.35.

   Draw two ways that he can pay the exact amount.

ball

2. Dakota has 2 dollar bills, 4 quarters, 2 dimes, and 1 nickel. She wants to buy a card that costs $2.75.

   Draw two ways she can pay the exact amount.

card

3. Kiran has 3 dollar bills, 1 quarter, 4 dimes, and 2 nickels. She wants to buy a kite that costs $3.50.

   Draw two ways that she can pay the exact amount.

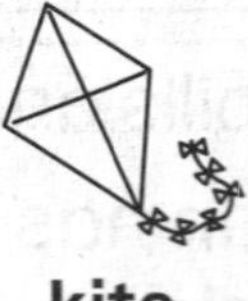
kite

Name ____________________ **Week 27**

Spiral Review

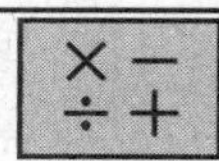

Use Workmat 6 and ●. Draw to show your work.
Write how many in each group.

1. Divide 18 ● into 2 equal groups.

_____ in each group

2. Divide 15 ● into 5 equal groups.

_____ in each group

3. Use the picture to complete the table.
Then shade the bars in the graph to show the data.

Circles	
Color	**Number**
black	
gray	
white	

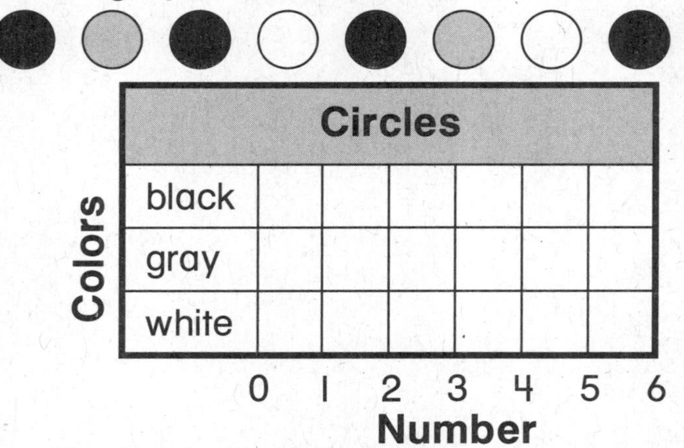

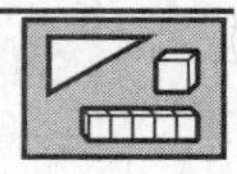

Use solid figures. Color the figure that matches the number of faces, edges, and vertices.

4. 5 faces, 8 edges, 5 vertices

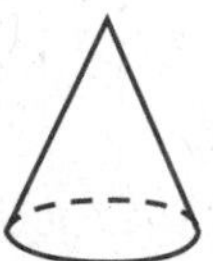 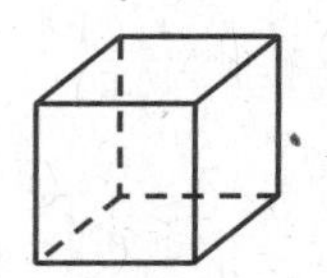 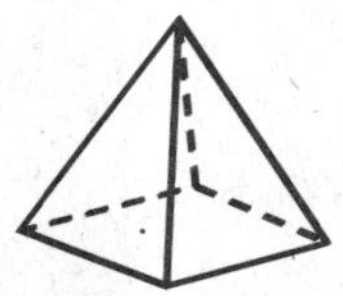 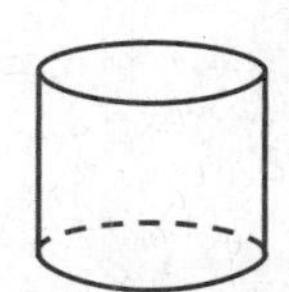

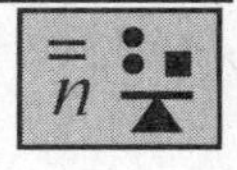

Cross out the information you do not need. Then solve.

5. There are 89 geese in a field.
There are 72 ducks and 95 geese in a lake. How many geese are there altogether?

_______ geese

Name____________________________________ **Lesson 17.3**

Compare Money Amounts

Who has the greater amount of money?

Zak's money

Keith's money

First, count the money. Write the amounts.

$1.12 | $1.30

Write a **dollar sign ($)** to the left of dollar amounts.

Use a **decimal point (.)** to separate dollars and cents.

Then, compare the amounts. Start with the dollars.

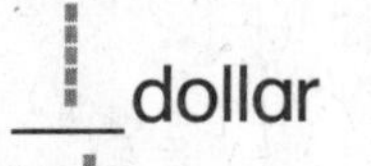

1 dollar  1 dollar

If the dollars are the same, compare the cents.

12 cents  30 cents

Keith has the greater amount of money.

Write each amount. Write who has the greater amount.

1. Rose's money

______

Liz's money

______

Who has the greater amount?

______

2. Sade's money

______

Olli's money

______

Who has the greater amount?

______

Name______________________________________ **Lesson 17.3**

Compare Money Amounts

Write each amount.
Then write who has the greater amount.

1. Leah's money

$1.20

Skip's money

____

Who has the greater amount?

____

2. Chuck's money

____

Tracy's money

____

Who has the greater amount?

____

3. C.J.'s money

____

Quiana's money

____

Who has the greater amount?

____

4. Dylan has 1 dollar bill, 2 dimes, and 2 pennies. Beata has 5 quarters. Katie has 1 dollar bill and 4 nickels. Who has the least amount of money? __________

Name ____________________

Lesson 17.4

Order Numbers

Compare the digits to put the numbers in order.

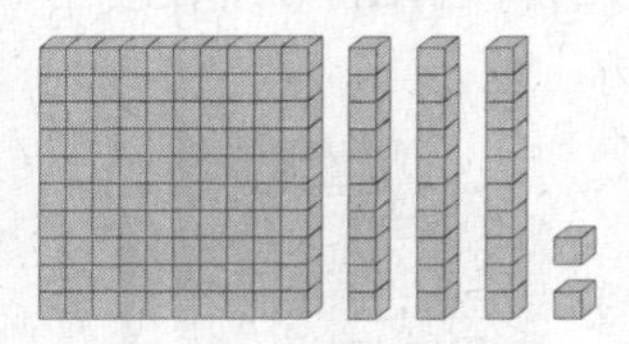

132

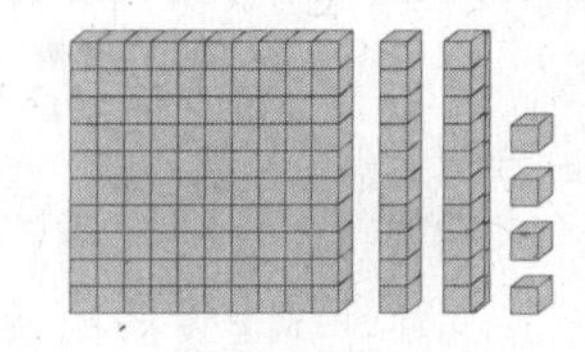

124

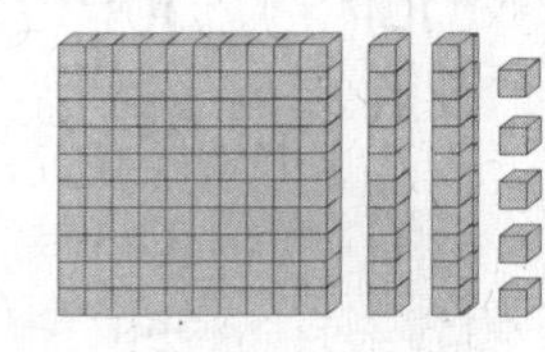

125

First, compare the hundreds.

The hundreds are the same.

Next, compare the tens.

132 has more tens than 124 and 125

Last, compare the ones.

125 has more ones than 124.

The greatest number is 132. The least number is 124.

132 (>) 125 (>) 124
greatest ... least

124 (<) 125 (<) 132
least ... greatest

Compare the numbers. Write them in the correct order. Then write > or < .

1.

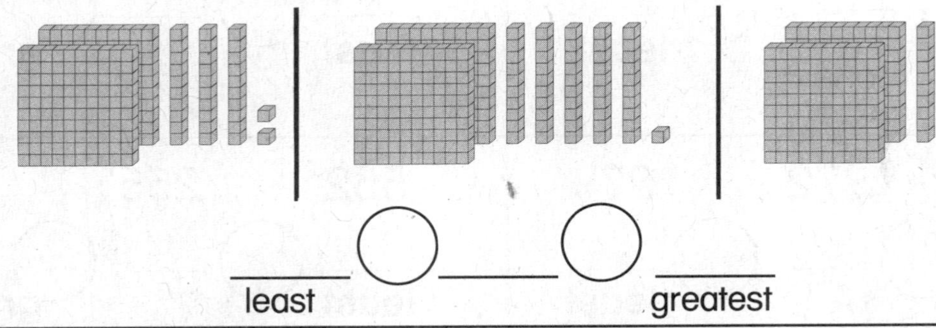

_____ ◯ _____ ◯ _____
least ... greatest

2.

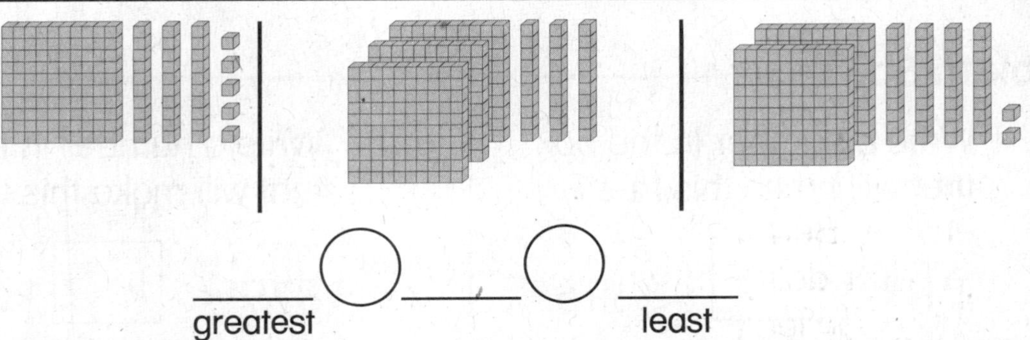

_____ ◯ _____ ◯ _____
greatest ... least

NS 1.3 Order and compare whole numbers to 1,000 by using the symbols <, =, >. **NS 1.0** Students understand the relationship between numbers, quantities, and place value in whole numbers up to 1,000.

Name____ **Lesson 17.4**

Order Numbers

Compare the numbers. Write them in the correct order. Then write > or <.

1.

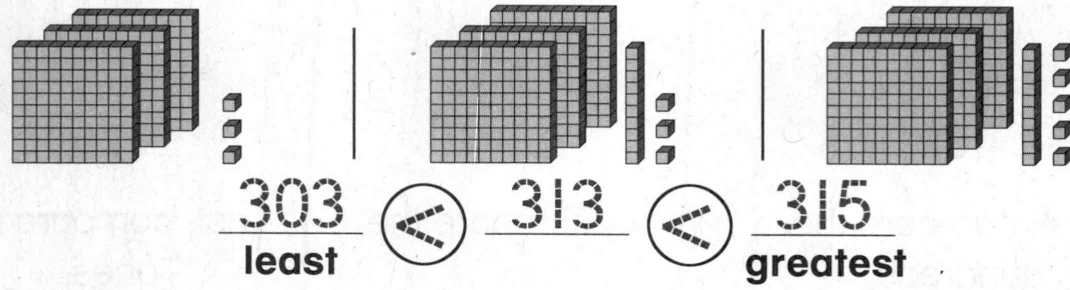

303 (<) 313 (<) 315

least — greatest

2.

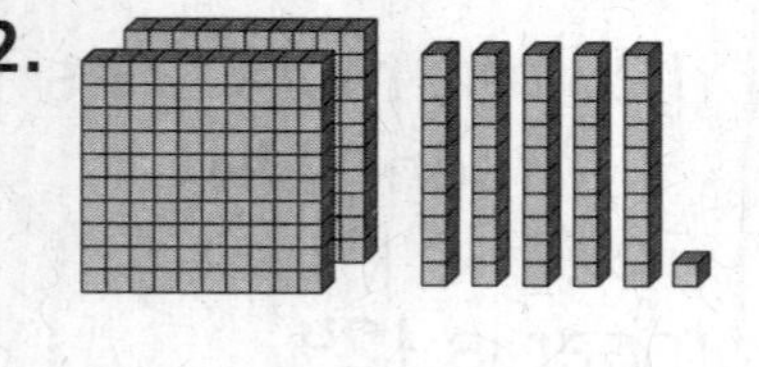

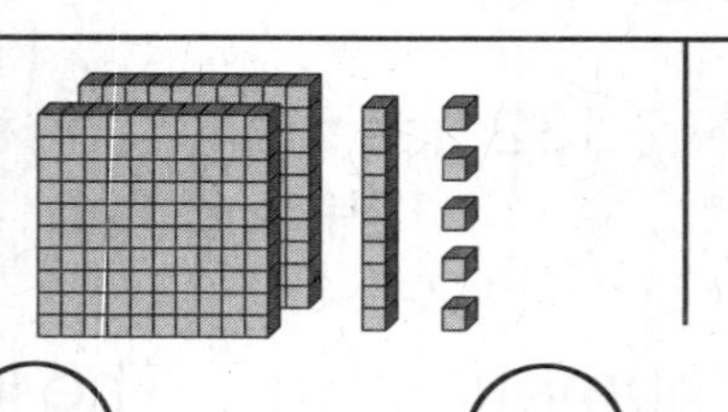

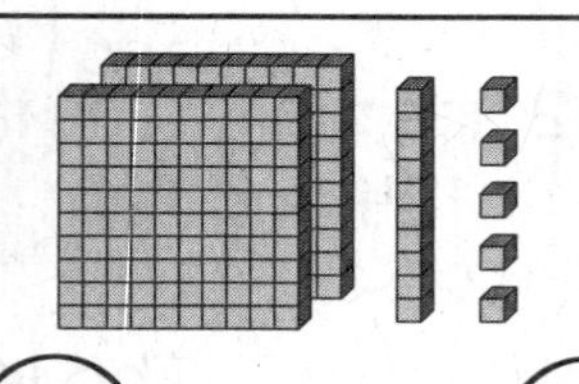

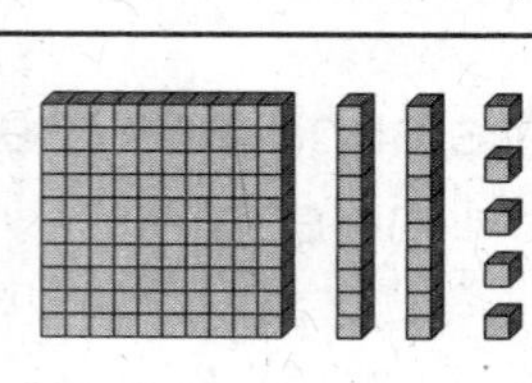

_____ ◯ _____ ◯ _____

least — greatest

Write the numbers in the correct order.
Then write > or < .

3. 166 351 407

_____ ◯ _____ ◯ _____

greatest — least

4. 740 760 750

_____ ◯ _____ ◯ _____

least — greatest

5. 873 972 274

_____ ◯ _____ ◯ _____

greatest — least

6. 552 255 525

_____ ◯ _____ ◯ _____

least — greatest

Problem Solving

7. Write a number in the box that will make this true.

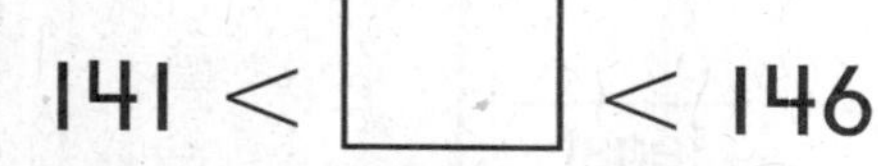

141 < ☐ < 146

8. Write a number in the box that will make this true.

879 < ☐ < 901

Spiral Review

Color the fraction strips to show the fractions.
Compare. Circle the greater fraction.

1. $\frac{1}{8}$ | | | | | | | | |

$\frac{1}{12}$ | | | | | | | | | | | | |

Find the pattern. Complete the table to solve.

2. How many wheels are on 5 wagons?

number of wagons	1	2	3	4	5
number of wheels	4	8	12		

There are ____ wheels on each wagon.

There are ________ wheels on 5 wagons.

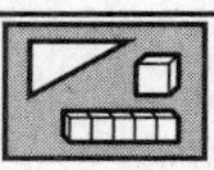

Use a clock. Write how much time has passed.

3. **Start**

4:00 P.M.

Finish

9:00 P.M.

____

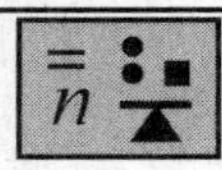

Find the sums.

4. $7 + 0 =$ ____

$4 + 0 =$ ____

5. $8 + 3 =$ ____

$8 + 4 =$ ____

6. $6 + 9 =$ ____

$9 + 6 =$ ____

Name ____________________

Practice 3-Digit Addition

Andy collects animal stamps. He has 135 polar bear stamps. He has 167 grizzly bear stamps. How many bear stamps does he have?

$$\begin{array}{r} 135 \\ +\ 167 \\ \hline \end{array}$$

Add the ones. Regroup 10 ones as 1 ten.	Add the tens. Regroup 10 tens as 1 hundred.	Add the hundreds.
$\begin{array}{r} ^{1} \\ 135 \\ +\ 167 \\ \hline 2 \end{array}$	$\begin{array}{r} ^{1\,1} \\ 135 \\ +\ 167 \\ \hline 02 \end{array}$	$\begin{array}{r} ^{1\,1} \\ 135 \\ +\ 167 \\ \hline 302 \end{array}$

There are 0 tens after regrouping. Use 0 as a place holder.

He has 302 bear stamps.

Add.

1. $\begin{array}{r} ^{1} \\ 345 \\ +\ 183 \\ \hline 528 \end{array}$	2. $\begin{array}{r} 687 \\ +\ 6 \\ \hline \end{array}$	3. $\begin{array}{r} 473 \\ +\ 18 \\ \hline \end{array}$	4. $\begin{array}{r} 291 \\ +\ 427 \\ \hline \end{array}$
5. $\begin{array}{r} 526 \\ +\ 57 \\ \hline \end{array}$	6. $\begin{array}{r} 438 \\ +\ 371 \\ \hline \end{array}$	7. $\begin{array}{r} 783 \\ +\ 8 \\ \hline \end{array}$	8. $\begin{array}{r} 656 \\ +\ 253 \\ \hline \end{array}$
9. $\begin{array}{r} 178 \\ +\ 119 \\ \hline \end{array}$	10. $\begin{array}{r} 462 \\ +\ 523 \\ \hline \end{array}$	11. $\begin{array}{r} 873 \\ +\ 39 \\ \hline \end{array}$	12. $\begin{array}{r} 582 \\ +\ 36 \\ \hline \end{array}$

NS 2.2 Find the sum or difference of two whole numbers up to three digits long. NS 2.0 Students estimate, calculate, and solve problems involving addition and subtraction of two- and three-digit numbers.

Name ____________________

Practice 3-Digit Addition

Add.

1. $\begin{array}{r} 1 \\ 218 \\ +\ 325 \\ \hline 543 \end{array}$	2. $\begin{array}{r} 641 \\ +\ 78 \\ \hline \end{array}$	3. $\begin{array}{r} 278 \\ +\ 15 \\ \hline \end{array}$	4. $\begin{array}{r} 492 \\ +\ 347 \\ \hline \end{array}$
5. $\begin{array}{r} 856 \\ +\ 8 \\ \hline \end{array}$	6. $\begin{array}{r} 318 \\ +\ 429 \\ \hline \end{array}$	7. $\begin{array}{r} 195 \\ +\ 513 \\ \hline \end{array}$	8. $\begin{array}{r} 374 \\ +\ 27 \\ \hline \end{array}$
9. $\begin{array}{r} 872 \\ +\ 36 \\ \hline \end{array}$	10. $\begin{array}{r} 608 \\ +\ 82 \\ \hline \end{array}$	11. $\begin{array}{r} 532 \\ +\ 235 \\ \hline \end{array}$	12. $\begin{array}{r} 795 \\ +\ 67 \\ \hline \end{array}$
13. $\begin{array}{r} 289 \\ +\ 55 \\ \hline \end{array}$	14. $\begin{array}{r} 377 \\ +\ 142 \\ \hline \end{array}$	15. $\begin{array}{r} 817 \\ +\ 4 \\ \hline \end{array}$	16. $\begin{array}{r} 541 \\ +\ 59 \\ \hline \end{array}$

Problem Solving

17. Chandra has 146 bear stickers. Daniela has double that many stickers. How many bear stickers do they have in all?

bear sticker

_______ bear stickers

Name

Problem Solving Skill: Too Much Information

Mr. Smith has 123 lion photos in a box. He has 94 lion photos in a photo album. He has 113 zebra photos in another box. How many lion photos does he have in all?

Photo

1. What are you asked to find out?
   How many lion photos Mr. Smith has in all.

2. Draw a line through the information you do not need to solve the problem.

Think: the question does not ask about zebra photos

3. Add to solve.

123	lion photos in a box
+ 94	lion photo in a photo album
217	lion photos in all

Mr. Smith has 217 lion photos in all.

Draw a line through any information you do not need Then solve.

4. There are 362 cows at Meyer Farm. There are 154 cows at Oak Farm. There are 238 cows at Kay Farm. How many cows are at Oak and Kay Farm together?

   _____ cows

5. There are 256 bluebirds in the trees. There are 98 blackbirds in the trees and 43 blackbirds on the ground. How many birds are in the trees?

   _____ birds

NS 2.0 Students estimate, calculate, and solve problems involving addition and subtraction of two- and three-digit numbers. **NS 2.2** Find the sum or difference of two whole numbers up to three digits long. **AF 1.0** Students, model, represent, and interpret number relationships to create and solve problems involving addition and subtraction. **AF 1.2** Relate problem situations to number sentences involving addition and subtraction.

Name ____________________

Skill • Too Much Information

Circle the information you need. Draw a line through the information you do not need. Solve.

1. There are 54 snow monkeys in the zoo. There are ~~38 lions~~ and 79 gibbon monkeys in the zoo. How many monkeys are in the zoo?

$$\begin{array}{r} 54 \\ +\ 79 \\ \hline 133 \end{array}$$

snow monkey

133 monkeys

---

2. There are 129 small dolphin toys and 145 large dolphin toys in a toy store. There are 112 small tiger toys. How many dolphin toys are in the store?

dolphin toy

______ dolphin toys

---

3. There are 156 sheep sleeping in the meadow. There are 118 sheep playing and 135 sheep eating. How many sheep are sleeping or eating?

sheep

______ sheep

---

4. A zoo worker feeds 92 seals on Monday. She feeds 85 sea lions and 38 otters on Tuesday. How many animals does she feed on Tuesday?

seal

______ animals

Spiral Review

Write how many.
Write the multiplication sentence.

1.

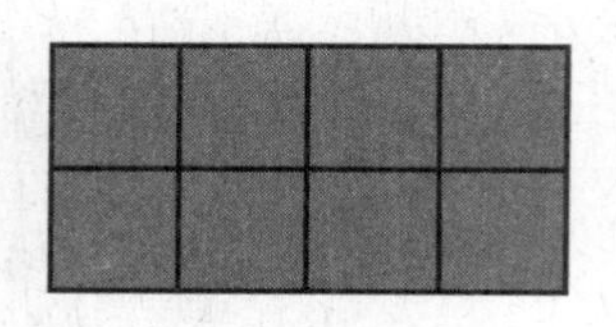

____ rows of ____

____ ◯ ____ ◯ ____

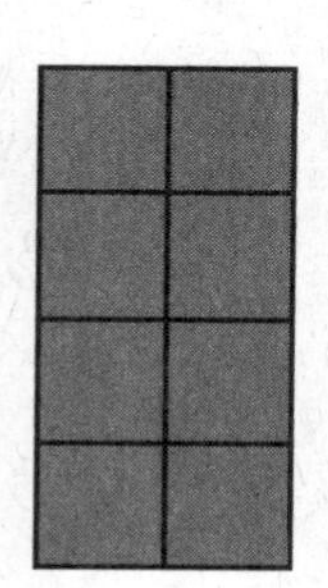

____ rows of ____

____ ◯ ____ ◯ ____

Use the table to answer the questions.

Nuts in a Bag	
Kind of Nut	**Number**
cashew	211
peanut	425
walnut	237
filbert	98
pecan	174

2. Of which kind of nut is there the most in the bag?

____

3. Of which kind of nut is there the least in the bag?

____

Measure the length to the nearest inch.

4.

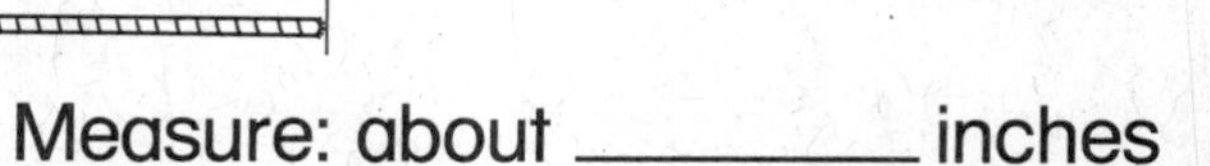

Measure: about ________ inches

Use the table to solve.

Fish in a Tank	
Color	**Number of Fish**
yellow	26
blue	43
orange	18
red	22

5. Mrs. Jagger puts 19 blue fish into a different tank. How many blue fish are left in the tank?

______ blue fish

Spiral Review

Write the numbers in the correct order.
Then write > or <.

1. 724 472 744

____ ◯ ____ ◯ ____
least greatest

2. 392 293 932

____ ◯ ____ ◯ ____
greatest least

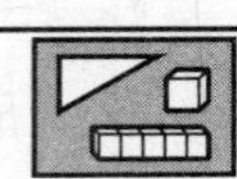

Use the hundred chart. Extend the pattern.

3. Skip-count by fives.

6, 11, 16, ____, ____, ____, ____, ____

4. Skip-count by twos.

7, 9, 11, ____, ____, ____, ____, ____

1	2	3	4	5	6	7	8	9	10
11	12	13	14	15	16	17	18	19	20
21	22	23	24	25	26	27	28	29	30
31	32	33	34	35	36	37	38	39	40
41	42	43	44	45	46	47	48	49	50

Use paper figures to make three new figures.
Cut along the folds. Trace the new figures.

5.

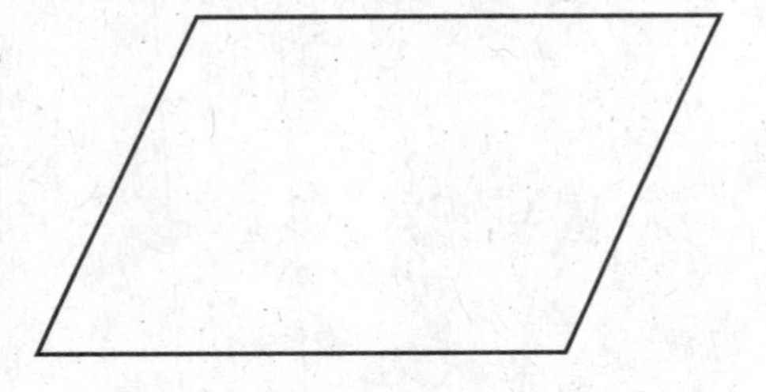

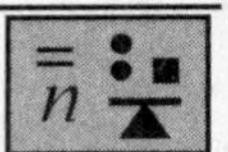

Draw a picture to solve.
Write a number sentence.

6. Tara has 6 blue beads and 9 orange beads. How many beads does she have in all?

______ beads

Name ______________________________

Spiral Review

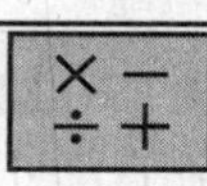

Use ●. Draw equal groups.
Write a multiplication sentence.

1. 3 groups of 4

4 + 4 + 4 = ____ ____ × ____ = ____

Skip-count. Write the missing numbers.
Write a rule for the pattern.

2. 42 43 44 45 46 47 48 49 50 51 52 53 54 55 56 57 58 59 60 61 62

42, 46, 50, ____, ____ Rule: Count by ________.

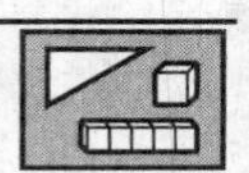

Use a cube. Measure the real object.

	Find the object.	Measure.
3.	chalkboard eraser	about ____ cubes

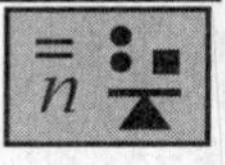

Write the sum.

4.	5.	6.	7.	8.
7	4	8	3	6
6	9	6	5	1
+ 1	+ 4	+ 2	+ 5	+ 4

Name____________________

Algebra: Multiply in Any Order

You can muliply numbers in any order.

This array shows 4 rows of 2 .

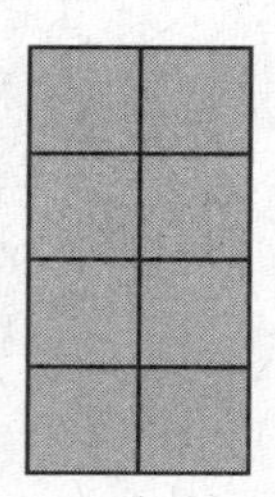

4 rows × 2 in each row = 8 in all

4 × 2 = 8

The **product** is 8.

This array shows 2 rows of 4 .

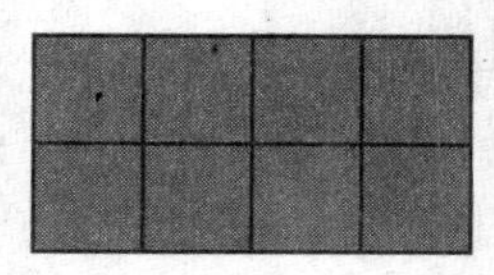

2 rows × 4 in each row = 8 in all

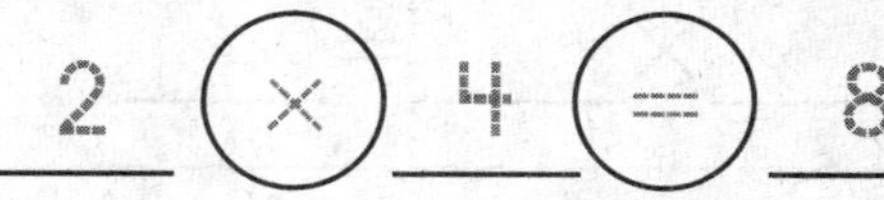

2 × 4 = 8

The **product** is 8.

Both problems multiply 2 and 4.
Both products are 8.
4 × 2 and 2 × 4 have the same product.

Write how many.
Write the multiplication sentence.

1.

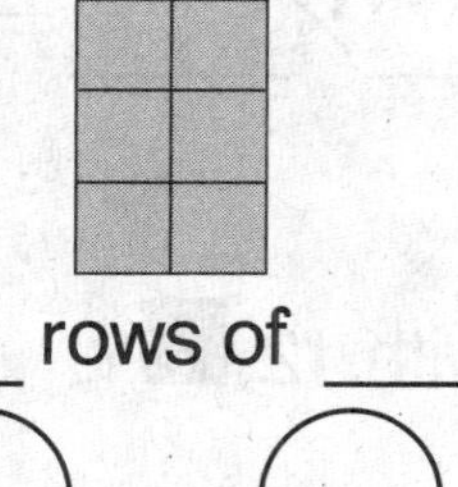

____ rows of ____

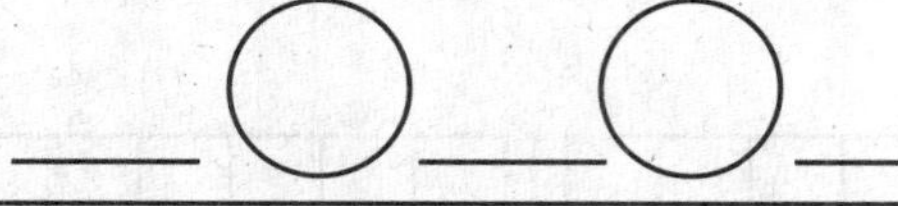

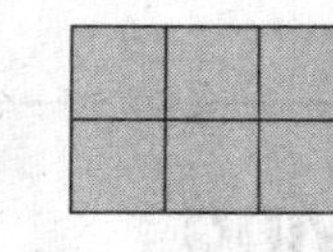

____ rows of ____

____ ◯ ____ ◯ ____

2.

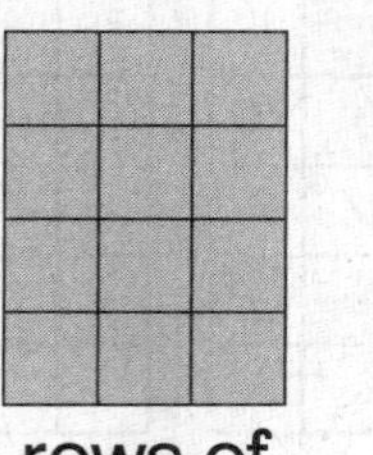

____ rows of ____

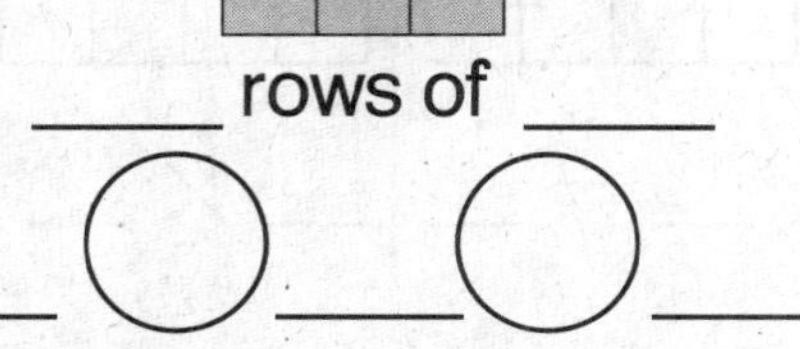

____ rows of ____

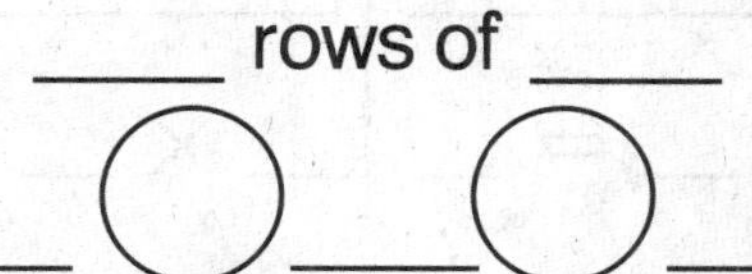

Name____ **Lesson 20.4**

Algebra: Multiply in Any Order

Multiply in any order.
The product is the same.

Write how many.
Write the multiplication sentence.

1.

2 rows of 4

____ × ____ = ____

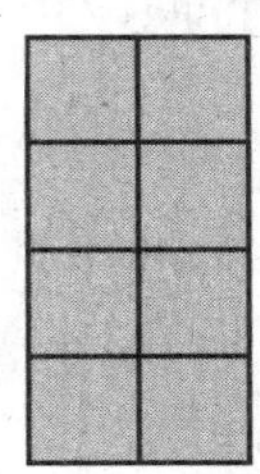

___ rows of ___

____ × ____ = ____

2.

___ rows of ___

____ × ____ = ____

___ rows of ___

____ × ____ = ____

Problem Solving

3. How many different arrays can you make with 12 ■?
Color the grids to show your work.

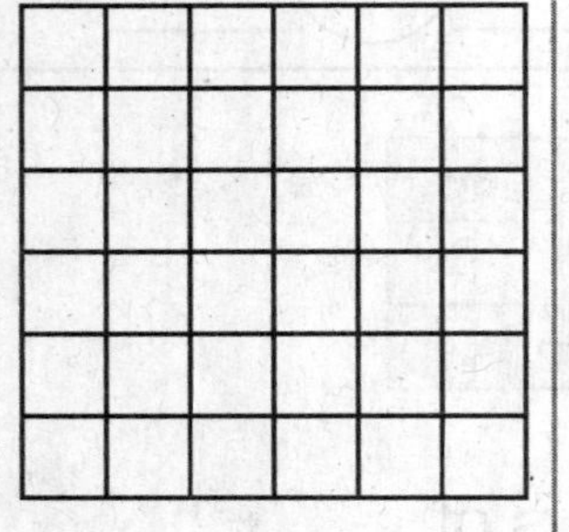

___ × ___ = ___

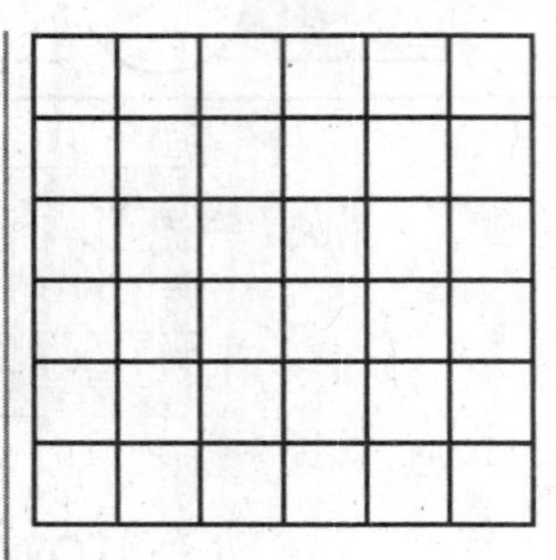

___ × ___ = ___

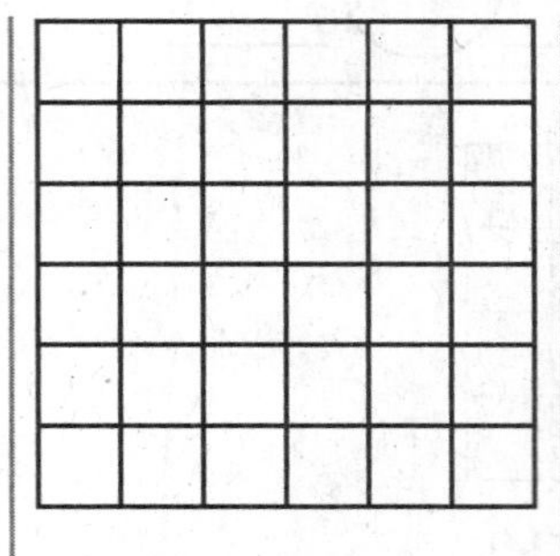

___ × ___ = ___

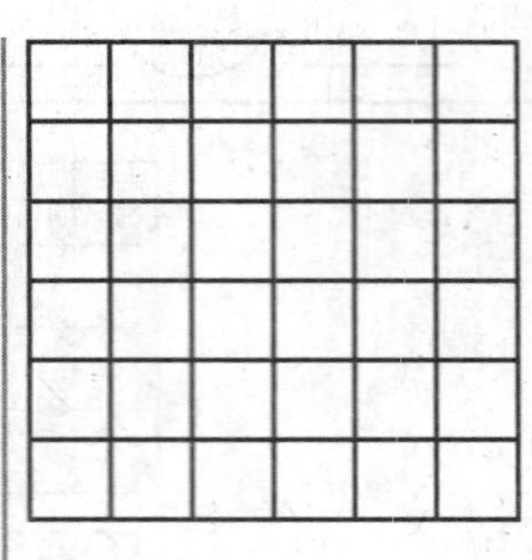

___ × ___ = ___

Name ____________________

Multiply with 2

There are 2 apples in each bag.
How many apples are in 5 bags?

Skip-count by twos to multiply with 2.
Skip-counting helps you find the product.

2,	4,	6,	8,	10,
There are 2 apples in 1 bag.	There are 4 apples in 2 bags.	There are 6 apples in 3 bags.	There are 8 apples in 4 bags.	There are 10 apples in 5 bags.
1 x 2 = 2	2 x 2 = 4	3 x 2 = 6	4 x 2 = 8	5 x 2 = 10

Skip-count by twos to find the product.

1.

4 x 2 = 8

2.

3 x 2 = ____

3.

1 x 2 = ____

4.

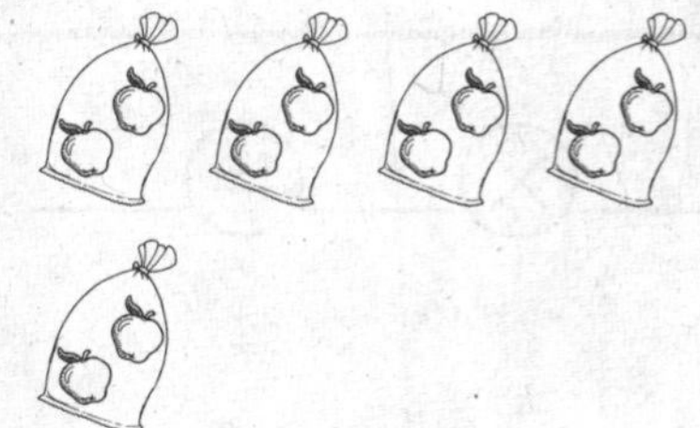

5 x 2 = ____

5.

2 x 2 = ____

6.

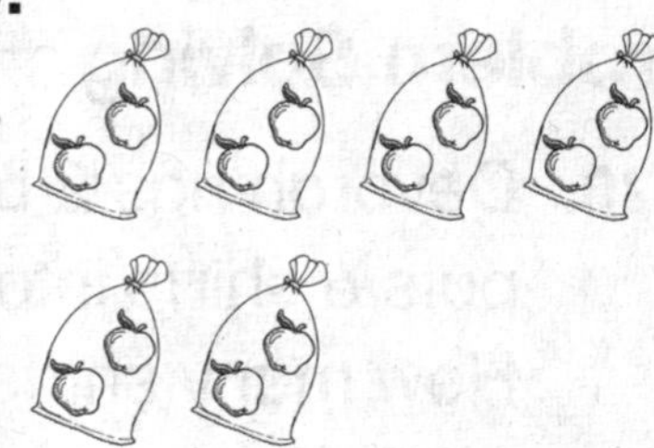

6 x 2 = ____

NS3.3 – Know the multiplication tables of 2s, 5s, and 10s (to "times 10") and commit them to memory. NS3.0 Students model and solve simple involving multiplication and division: NS3.1 Use repeated addition, arrays, and counting by multiples to do multiplication.

Name ______________________________

Multiply with 2

Skip-count by twos to find the product.

1. $8 \times 2 =$ 16

2. $7 \times 2 =$ ____

3. 

$9 \times 2 =$ ____

Write the product.

You can multiply in any order. The product is the same

4.	5.	6.	7.	8.	9.
5×2	2×4	2×7	2×3	6×2	3×2

10.	11.	12.	13.	14.	15.
2×6	2×8	2×5	2×9	2×7	4×2

16.	17.	18.	19.
2×6 6×2	7×2 2×7	2×5 5×2	2×8 8×2

Problem Solving

20. Devorah has 2 baskets. She puts 6 shirts into each basket. How many shirts does she put in the baskets?

shirt

____ $\times$ ____ $=$ ____

____ shirts

Name ______________________________ **Week 32**

Spiral Review

Use Workmat 6 and ●.
Draw to show your work. Write how many groups and the remainder.

1. Divide 14 ● into groups of 3.

_____ groups

The remainder is _____.

2. Divide 9 ● into groups of 4.

_____ groups

The remainder is _____.

Write a title and labels for the bar graph. Then complete the bar graph to show the information.

3. Nikki took a survey of her classmates' favorite lunches. 5 classmates chose pizza, 6 chose tacos, 3 chose chili, and 2 chose pasta.

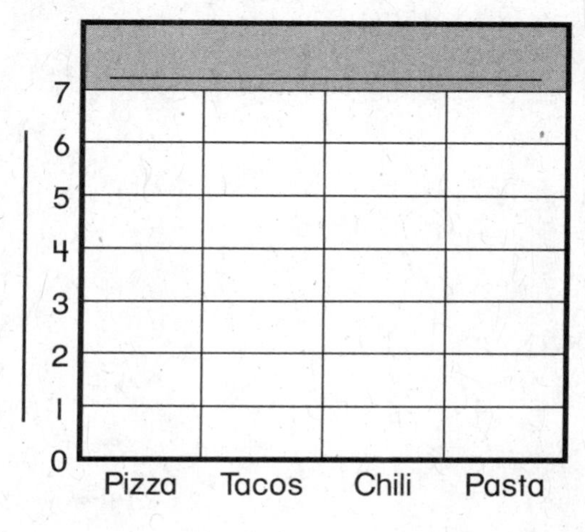

Write a title to describe the group of plane figures.

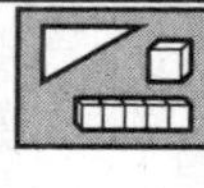

4. ______________________________

____

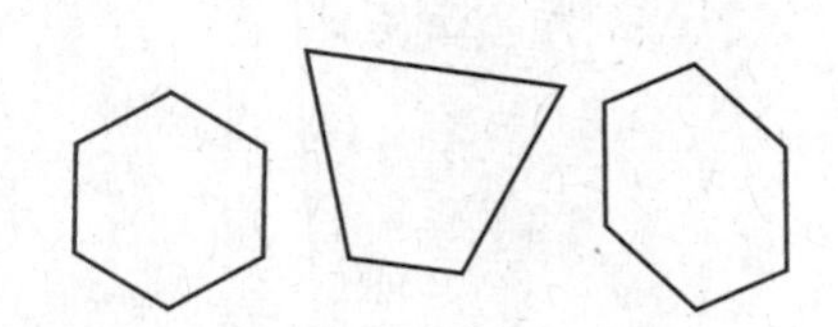

Use Workmat 5 and ▦ ▬ ▪. Subtract.

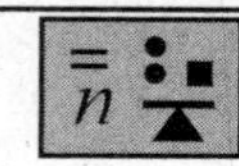

5.

Hundreds	Tens	Ones
☐	☐	☐
7	3	4
− 3	2	5

Name________________________________

Lesson 21.1

Size of Shares

When you divide, you make equal groups.
Divide 6 ● into 2 equal groups.
How many ● will be in each group?

First, show 2 groups and 6 ●.

Put one ● in each group.

Put another ● in each group.

Put one more ● in each group.

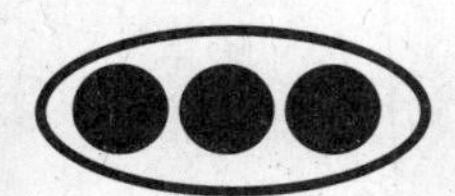 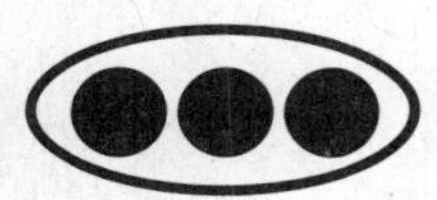

All 6 ● have been grouped.

There are 3 in each group.

Use Workmat 6 and ●. Draw to show your work.
Write how many in each group.

1. Divide 12 ● into 2 equal groups.

_____ in each group

2. Divide 9 ● into 3 equal groups.

_____ in each group

NS3.2 Use repeated subtraction, equal sharing, and forming equal groups with remainders to do division. NS 3.0 Students model and solve simple problems involving multiplication and division:

Name ____________________ **Lesson 21.1**

Size of Shares

Use Workmat 6 and ●. Draw to show your work.
Write how many in each group.

1. Divide 12 ● into 3 equal groups.

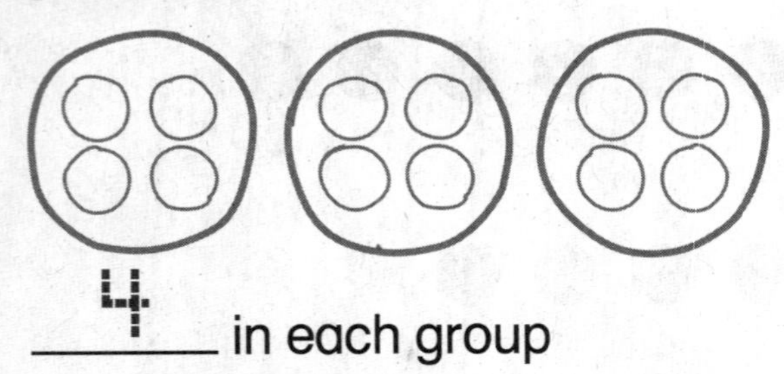

_______ in each group

2. Divide 8 ● into 4 equal groups.

_______ in each group

3. Divide 10 ● into 2 equal groups.

_______ in each group

4. Divide 14 ● into 2 equal groups.

_______ in each group

5. Divide 18 ● into 3 equal groups.

_______ in each group

6. Divide 15 ● into 5 equal groups.

_______ in each group

Problem Solving

7. Mr. Carr divides 6 meatballs into 3 bowls.
He wants to have 3 meatballs in each bowl.
How many more meatballs does he need?

_______ more meatballs

Name____

Number of Equal Shares

When you **divide**, make **equal** groups.
Then count how many groups you made.

Divide 6 ● into groups of 2.
How many groups will you make?

● ● ● ● ● ●

Show 6 ●. Make a group of 2 ●.

● ● ● ●

Make another group of 2 ●.

● ●

Make another group of 2 ●.

All 6 ● have been grouped.
There are 3 groups.

6 ÷ 2 = 3 is a division sentence.

It tells about dividing 6 ● into groups of 2.

Use Workmat 6 and ●. Draw to show your work.
Write a division sentence.

1. Divide 12 ● into groups of 4.

___ ◯ ___ ◯ ___

_______ groups

2. Divide 16 ● into groups of 8.

___ ◯ ___ ◯ ___

_______ groups

NS3.2 Use repeated subtraction, equal sharing, and forming equal groups with remainders to do division. NS 3.0 Students model and solve simple problems involving multiplication and division:

Name________________________________

Number of Equal Shares

Use ●. Draw to show your work.

Write how many equal groups.

1. Divide 8 ● into groups of 2.

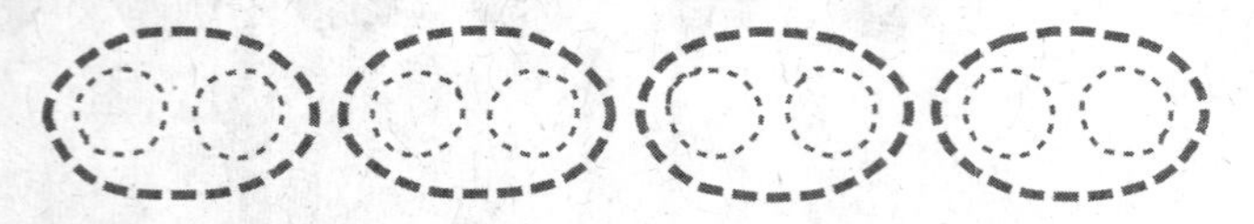

4 groups

2. Divide 15 ● into groups of 5.

_____ groups

3. Divide 14 ● into groups of 2.

_____ groups

4. Divide 20 ● into groups of 4.

_____ groups

Problem Solving

Use ●. Draw to show your work.

5. Mrs. Rodriguez takes 5 children to the fair. She has $12. She wants to give each child some money for rides. Can she give each child $3?

____

Name____

Equal Shares with Remainders

Divide 7 ● into groups of 3.

First, show 7 ●.

Make a group of 3.

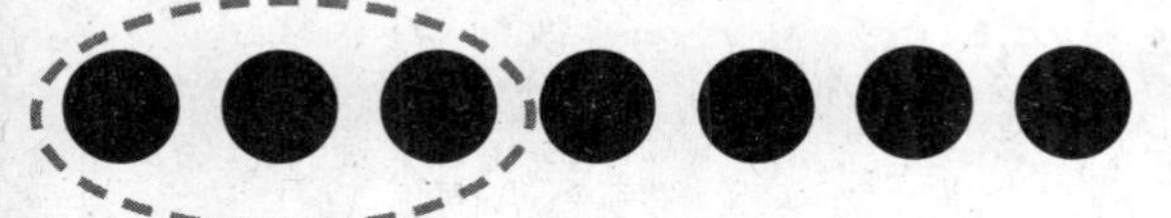

Make another group of 3.

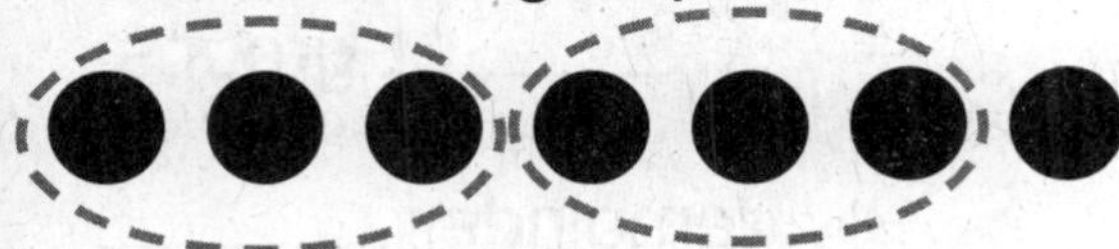

Think:
There are fewer than 3 dots left. You cannot make another group of 3.

You made 2 groups of 3.

There is __1__ left over.

This is the remainder.

There are __2__ groups. The remainder is __1__.

Use Workmat 6 and ●. Draw to show your work.
Write how many groups and the remainder.

1. Divide 9 ● into groups of 4.

_____ groups

The remainder is _____.

2. Divide 11 ● into groups of 3.

_____ groups

The remainder is _____.

NS3.2 Use repeated subtraction, equal sharing, and forming equal groups with remainders to do division. **NS 3.0** Students model and solve simple problems involving multiplication and division:

Name________________________________

Equal Shares with Remainders

Use Workmat 6 and ●. Draw to show your work.
Write how many groups and the remainder.

1. Divide 10 ● into groups of 4.

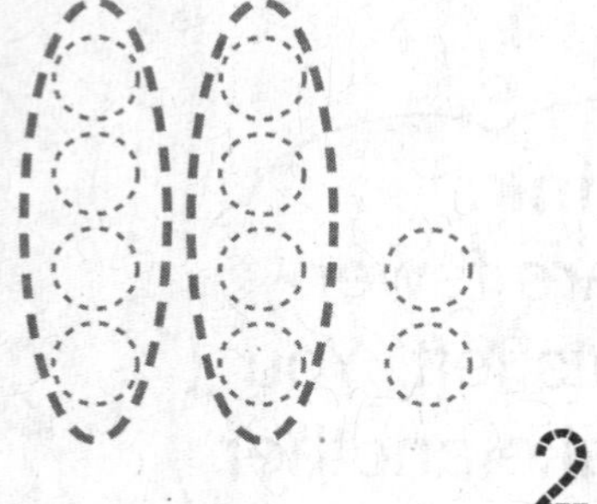

2 groups

remainder 2

2. Divide 13 ● into groups of 3.

_______ groups

remainder _______

3. Divide 12 ● into groups of 6.

_______ groups

remainder _______

4. Divide 15 ● into groups of 4.

_______ groups

remainder _______

Problem Solving

5. Barry has 11 toy cars. He gives some of his friends 3 cars each. He has 2 cars left over. How many friends get cars?

_______ friends

Name ______________________ **Week 33**

Spiral Review

Use Workmat 6 and ●. Draw to show your work.
Write a division sentence.

1. Divide 16 ● into groups of 2.

___ ◯ ___ ◯ ___

___ groups

2. Divide 10 ● into groups of 5.

___ ◯ ___ ◯ ___

___ groups

Find the pattern. Complete the table to solve.

3. How many legs are on 5 beetles?

number of beetles	1	2	3	4	5
number of legs	6	12	18		

There are ___ legs on each beetle.

There are ___ legs on 5 beetles.

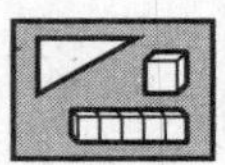

Measure the length to the nearest centimeter.

4.

Measure: about ___ centimeters

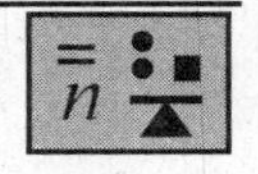

Draw a picture to solve.
Write a number sentence.

5. There are 5 flowers in a vase. Each flower has 4 petals. How many petals are there in all?

___ petals

Name________________________________ Lesson 22.1

Measure Length with Nonstandard Units

You can use cubes to measure length.
Line up the cubes and count them.

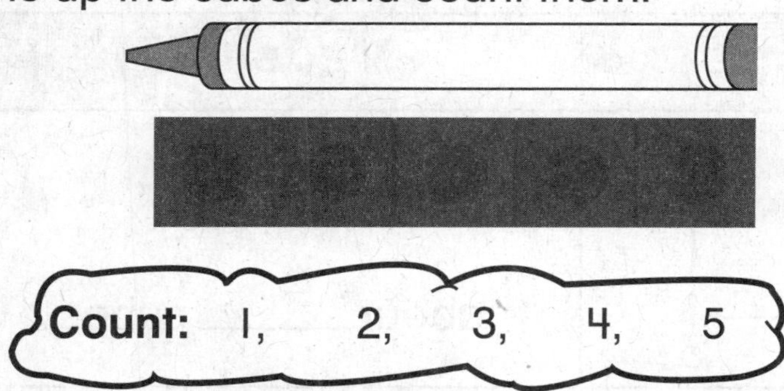

Line up the first cube with one end of the object. Keep until you reach the other end.

Count: 1, 2, 3, 4, 5

This crayon is about 5 cubes long.

Use [cube]. Measure the real object.

	Find the object.	Measure.
1.	marker	about ____ cubes
2.	bookshelf	about ____ cubes
3.	card	about ____ cubes

MG 1.1 Measure the length of objects by iterating (repeating) a nonstandard or standard unit. MG 1.0 Students understand that measurement is accomplished by identifying a unit of measure, iterating (repeating) that unit, and comparing it to the item to be measured.

Name____

Measure Length with Nonstandard Units

Use [cube]. Measure the real object.

	Find the object.	Measure.
1.	pencil	about _____ cubes
2.	folder	about _____ cubes
3.	tape	about _____ cubes
4.	glue	about _____ cubes

Problem Solving

Choose two objects. Estimate the length of each by comparing them to the objects you have just measured.

5. Object: ________________

Estimate: about _____ cubes

Measure: about _____ cubes

6. Object: ________________

Estimate: about _____ cubes

Measure: about _____ cubes

Name ______________________

Compare Nonstandard Units

You can measure length with different units. Use larger units to measure the length of longer objects.

Straws are a better unit for measuring the blackboard.

Use smaller units to measure the the length of shorter objects.

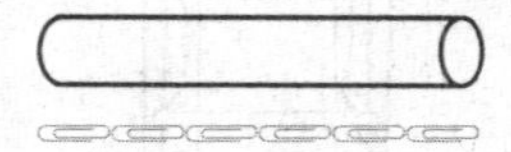

Paper clips are a better unit for measuring the chalk.

Measure the length of the real object. Choose the better unit to use.

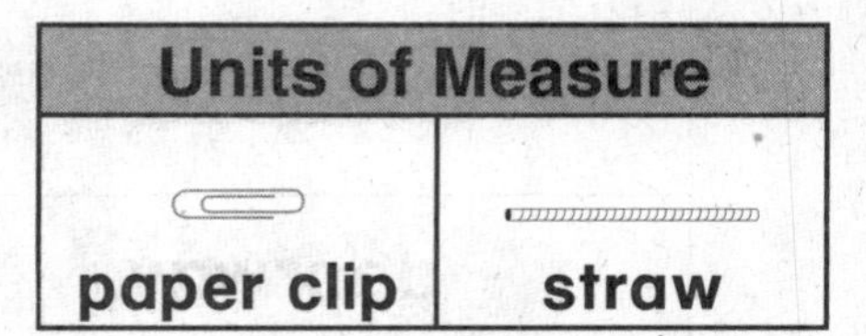

Units of Measure	
paper clip	straw

	Find the object.	Choose the unit.	Measure.
1.	paper	____________	about ____________
2.	wall	____________	about ____________

O—π **MG 1.2** Use different units to measure the same object and predict whether the measure will be greater or smaller when a different unit is used. **MG 1.1** Measure the length of objects by iterating (repeating) a nonstandard or standard unit. **MG 1.0** Students understand that measurement is accomplished by identifying a unit of measure, iterating (repeating) that unit, and comparing it to the item to be measured.

Name ________________________________

Compare Nonstandard Units

Measure the length of the real object with paper clips and with straws.

Units of Measure	
paper clip	straw

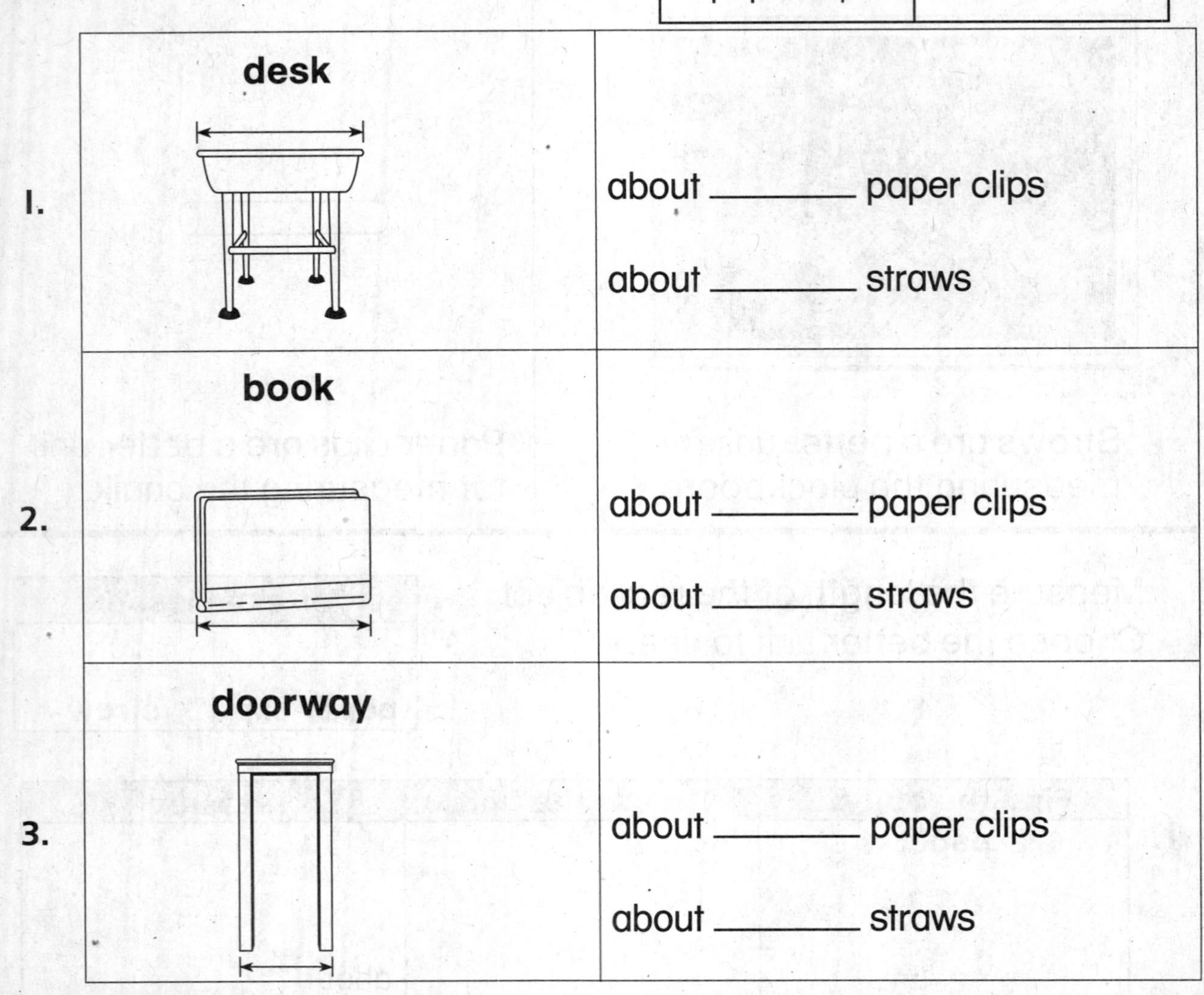

1. desk	about ________ paper clips about ________ straws
2. book	about ________ paper clips about ________ straws
3. doorway	about ________ paper clips about ________ straws

Problem Solving

4. This rope is about 12 beads long.
   About how many paper clips long is the rope?

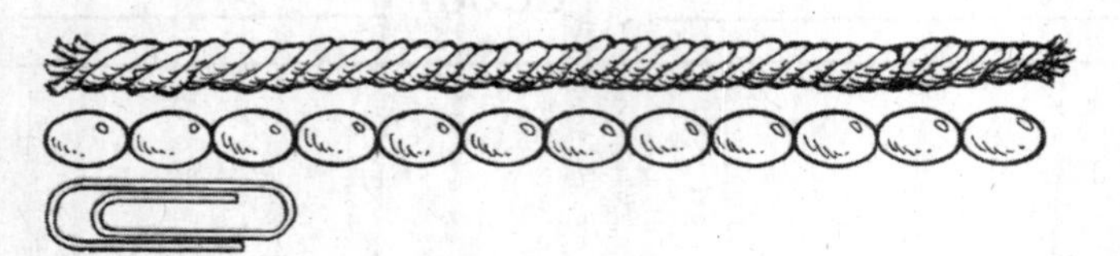

about ______ paper clips

Spiral Review

Circle to show equal parts. Then write if the black part is one **half**, **third**, or **fourth** of the group.

1. 3 equal parts

One ______________ of the group is black.

2. 2 equal parts

One ______________ of the group is black.

Use the line plot.

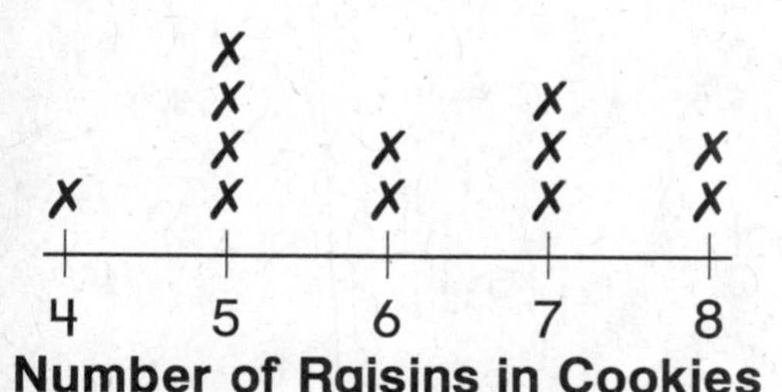

3. Write a number sentence to find the range of the data.

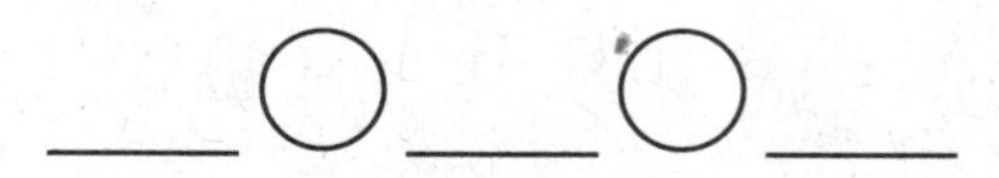

Write **more than**, **less than**, or **the same as** to complete the sentence.

Time Relationships
There are 24 hours in 1 day.
There are 7 days in 1 week.
There are about 4 weeks in 1 month.
There are 12 months in 1 year.
There are about 52 weeks in 1 year.

4. It snowed for 5 days in the mountains.

This is ______________ 1 week.

Use the picture graph.

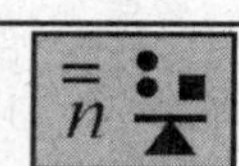

Trees in the Park						
elm	○	○	○	○	○	
oak	○	○				
maple	○	○	○			

Each ○ stands for 1 tree.

5. How many oak trees are in the park?

_____ oak trees

6. How many trees are in the park altogether?

_____ trees

Name________________________________

Lesson 22.5

Measure in Inches and Feet

Use inches to measure smaller objects.

A stamp is about I **inch** long.

Use feet to measure larger objects.

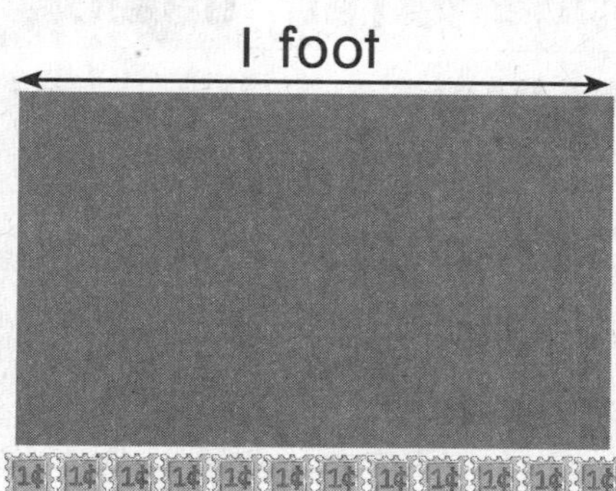

I 2 stamps are about I foot long.

A folder is about I **foot** long.

Measure the length of the object.

	Find the real object.	Measure.
I.	crayon (RED)	about ______ inches
2.	bookshelf	about ______ feet

Measure in inches. Then measure in feet.

	Find the real object.	Measure.
3.	rug	about ______ inches about ______ feet

MG 1.2 – Use different units to measure the same object and predict whether the measure will be greater or smaller when a different unit is used. **MG 1.3** Measure the length of an object to the nearest inch and/or centimeter. **NS 6.1** Recognize when an estimate is reasonable in measurements (e.g., closest inch).

Name________________________________

Lesson 22.5

Measure in Inches and Feet

Measure in inches. Then measure in feet.

	Find the real object.	Measure.
1.	bulletin board	about ________ inches about ________ feet
2.	teacher's desk	about ________ inches about ________ feet

Circle the better estimate.

3. **stapler**

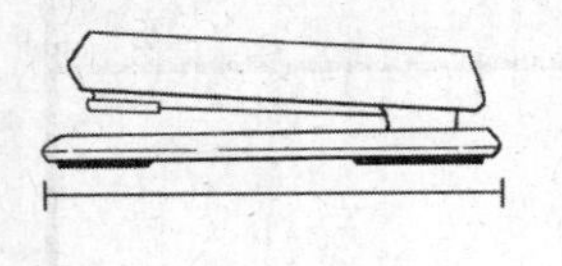

about 7 inches

about 7 feet

4. **poster**

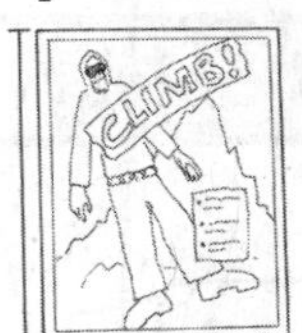

about 3 inches

about 3 feet

5. **desk**

about 2 feet

about 2 inches

6. **chalk**

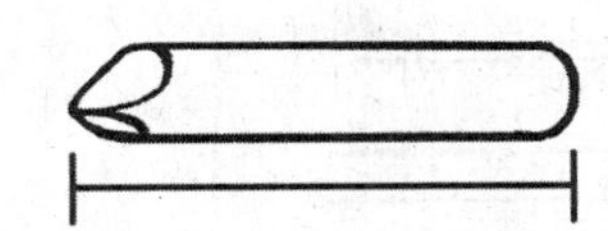

about 4 feet

about 4 inches

Problem Solving

Circle the best estimate.

7. About how long is a pencil?

about 6 inches

about 16 inches

about 6 feet

8. About how tall is a door?

about 17 feet

about 7 feet

about 7 inches

Name ______________________________________ **Lesson 23.3**

Cups, Pints, Quarts, and Gallons

The **capacity** of a container is how much the container holds. There are different ways to measure capacity.

A **cup** is one of the smallest measurements.

It takes 2 cups to make 1 **pint**.

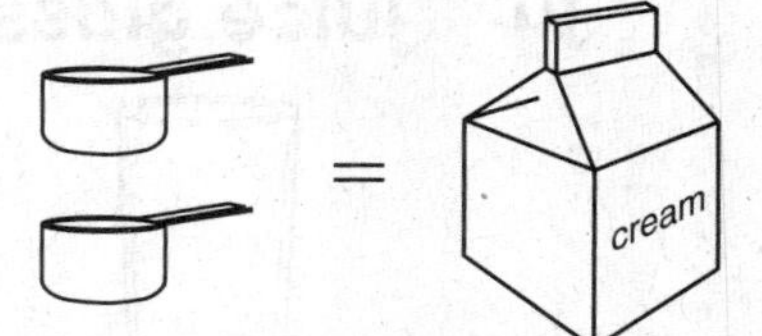

It takes 4 cups to make 1 **quart**.

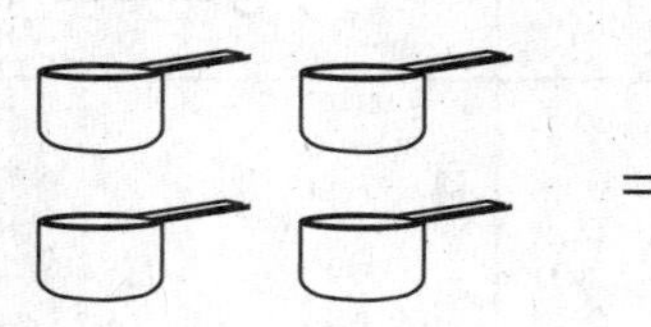

=

There are 2 pints in 1 quart.

It takes 16 cups to make 1 **gallon**.

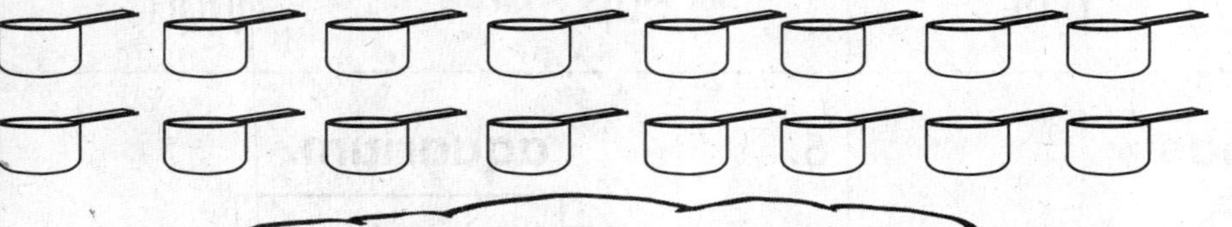

=

There are 4 quarts in 1 gallon.
There are 8 pints in 1 gallon.

How many cups does the container hold?
How many pints does the container hold?

1.

soy milk

_______ cups

_______ pints

2.

juice
1 pint

_______ cups

_______ pints

3.

_______ cups

_______ pints

4.

_______ cups

_______ pints

MG 1.0 Students understand that measurement is accomplished by identifying a unit of measure, iterating (repeating) that unit, and comparing it to the item to be measured. **MG 1.2** Use different units to measure the same object and predict whether the measure will be greater or smaller when a different unit is used.

Name____

Hands On: Cups, Pints, Quarts, and Gallons

Circle the better unit of measure for the capacity of the container.

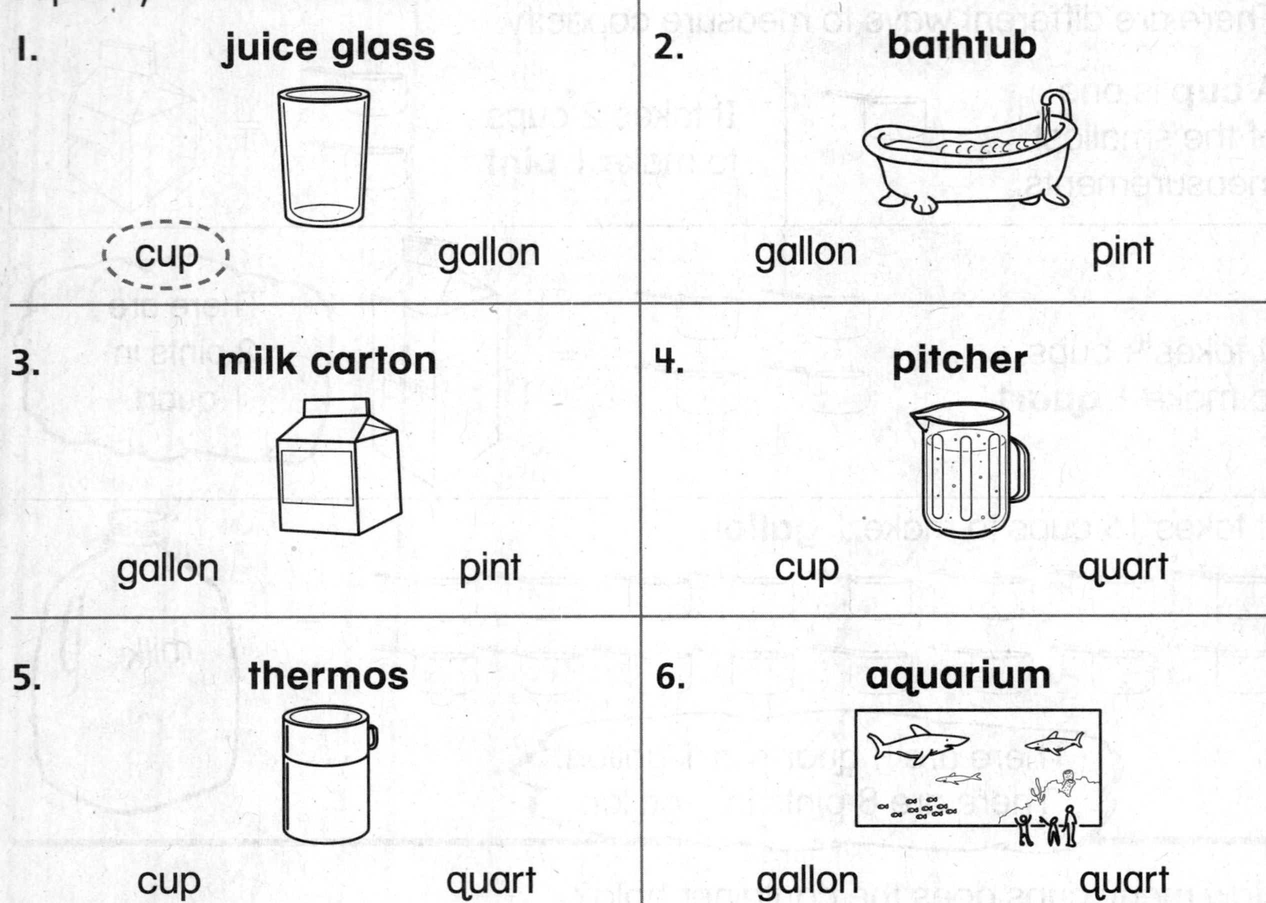

1. **juice glass**
   cup gallon

2. **bathtub**
   gallon pint

3. **milk carton**
   gallon pint

4. **pitcher**
   cup quart

5. **thermos**
   cup quart

6. **aquarium**
   gallon quart

Problem Solving

Circle the better answer.

7. Abby is using a 1 cup container to fill the vase with water. The vase holds 8 cups. She will only fill the vase halfway. How much water will be in the vase when Abby is finished?

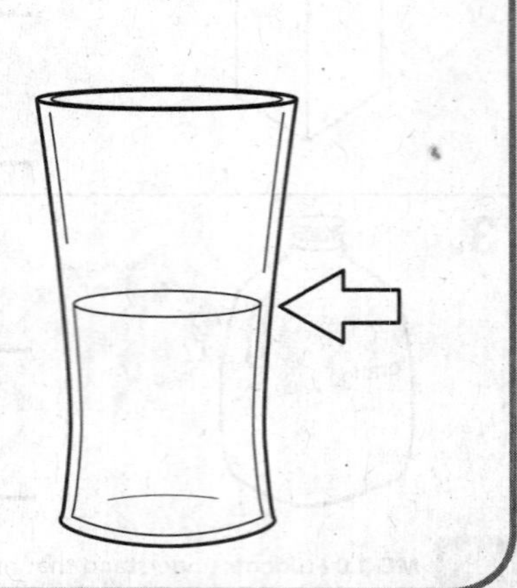

about 1 pint about 1 quart

Name ______________________ **Week 35**

Spiral Review

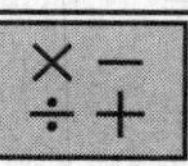

Use ■. Color the array.
Write a multiplication sentence.

1. 6 rows of 3

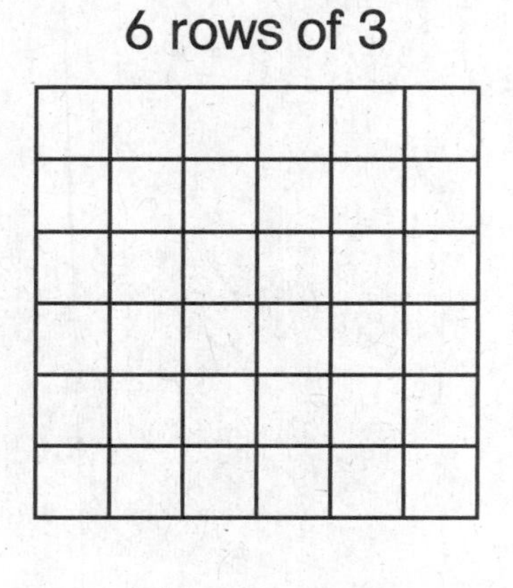

_____ × _____ = _____

2. 2 rows of 5

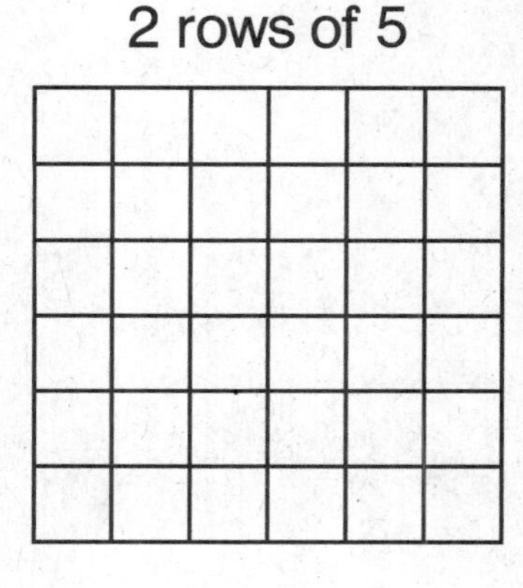

_____ × _____ = _____

Skip-count. Write the missing numbers.
Write a rule for the pattern.

3. 52 53 54 55 56 57 58 59 60 61 62 63 64 65 66 67 68 69 70 71 72

52, 55, 58, _____, _____ Rule: Count by __________.

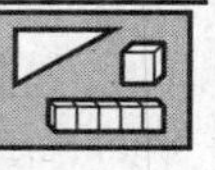

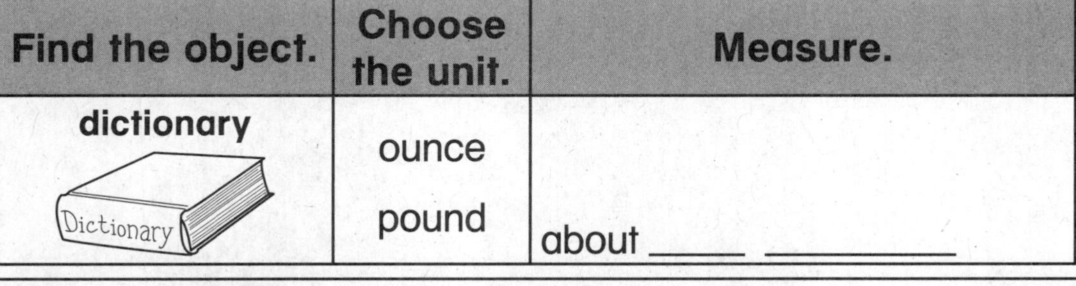

	Find the object.	Choose the unit.	Measure.
4.	**dictionary**	ounce pound	about _____ _________

Use Workmat 11 and to make a model.
Write how many.

5. Freddy has 17 toy trains on the floor in his bedroom. He puts 9 trains in a toy chest. How many trains are left on the floor?

_____ trains

Name______________________________________

Liters

You can measure capacity in liters.
One liter is about the same amount as one quart.

This container holds 1 **liter**.

This jug of milk holds more than 1 liter.

This carton of milk holds less than 1 liter.

Estimate how many liters the container can hold.
Then measure.

1. **recycling bin**

Estimate: about ________ liters

Measure: about ________ liters

2. **detergent bottle**

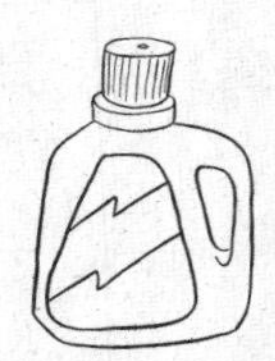

Estimate: about ________ liters

Measure: about ________ liters

3. **sink**

Estimate: about ________ liters

Measure: about ________ liters

MG 1.0 Students understand that measurement is accomplished by identifying a unit of measure, iterating (repeating) that unit, and comparing it to the item to be measured.

Name

Hands On: Liters

Estimate how much the container can hold.

Circle **more than 1 liter** or **less than 1 liter.**

1. **ladle**  more than 1 liter / less than 1 liter	2. **bucket**  more than 1 liter / less than 1 liter	3. **measuring cup**  more than 1 liter / less than 1 liter
4. **soap bottle**  more than 1 liter / less than 1 liter	5. **soup can**  more than 1 liter / less than 1 liter	6. **sink**  more than 1 liter / less than 1 liter

Problem Solving

7. Think about some containers in your classroom. What container might hold more than 1 liter? Draw and label the container.

Name____

Problem Solving Skill: Make Reasonable Estimates

Circle the better estimate.

Rex is drinking some juice. About how much juice might he have?

about I gallon

about I pint

I. What are you asked to find out?

how much juice Rex has

2. Think about how much I pint and I gallon mean in cups.

I pint = 2 cups   I gallon = 16 cups

3. Which is a more reasonable amount for Rex to drink?

16 cups   (2 cups)

4. Circle the better estimate.

about I gallon   (about I pint)

Circle the better estimate.

5. Carolina buys some milk at the grocery store. About how much milk might Carolina buy?

about I gallon

about 10 gallons

6. Julio picks up his new baseball bat. About how much might the bat weigh?

about 2 ounces

about 2 pounds

Name ______________________________

Skill • Make Reasonable Estimates

Circle the better estimate.

1. Fiona is pouring a glass of milk. About how much milk might she pour?

about 2 cups

about 20 cups

2. Charlie is holding a basketball. What might the weight of the basketball be?

about 1 ounce

about 1 pound

3. Molly wants to fill the bathtub with water. About how much water might it hold?

about 1 liter

about 50 liters

4. Henry picks up his trumpet to begin playing. What might the mass of the trumpet be?

about 2 grams

about 2 kilograms

5. Adele makes a pitcher of fruit punch for her friends. About how much might the pitcher hold?

about 2 quarts

about 20 quarts

Name________________________________

Time to 15 Minutes

On the hour, the minute hand points to the 12.

After 15 minutes, the minute hand has moved one quarter of the way around. It points to the 3. The time is 10:15.

What is the time after 15 more minutes?

The time is 10:30.

This is 30 minutes after 10 o'clock.

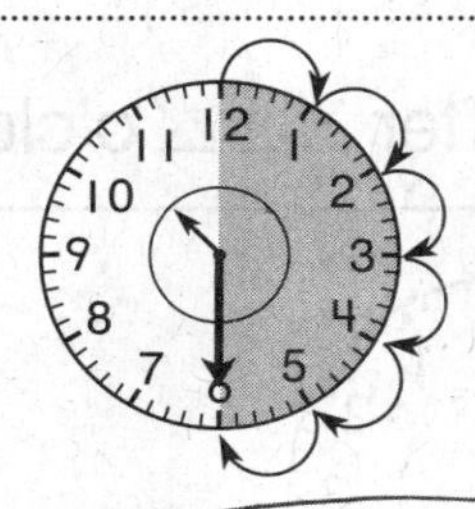

Write the time in two ways.

REMEMBER: count by 5s.

1.

30 minutes after 2 o'clock

2.

____ minutes after ____ o'clock

3.

____ minutes after ____ o'clock

4.

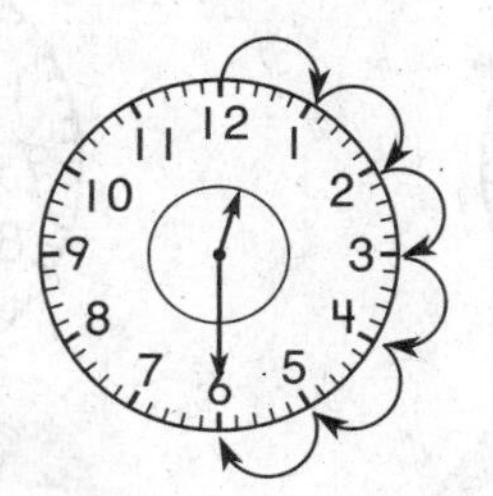

____ minutes after ____ o'clock

Name____

Lesson 24.2

Time to 15 Minutes

REMEMBER:
Count by fives.

Write the time in two ways.

1.

30 minutes after 1 o'clock

2.

_____ minutes after _____ o'clock

3.

_____ minutes after _____ o'clock

4.

_____ minutes after _____ o'clock

Problem Solving

Look at where the hour hand is pointing. About what time is it?

5.

6.

7.

Spiral Review

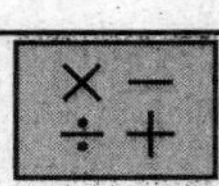

Use a number line to write a division sentence.

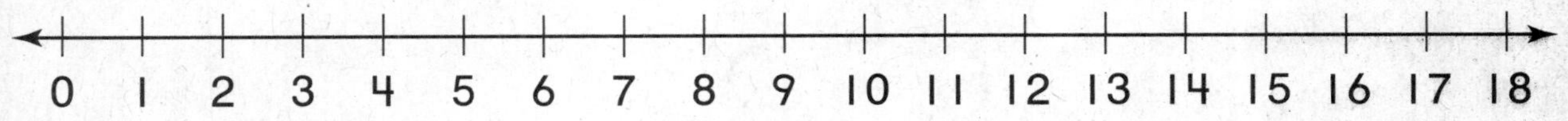

1. Divide 16 into equal groups of 2.

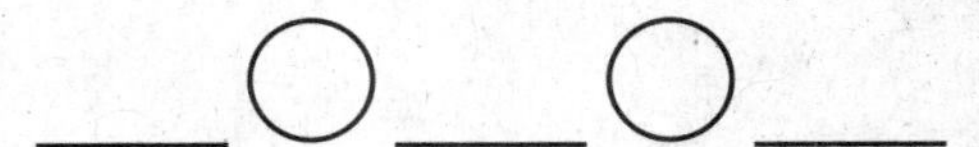

Use the bar graph.

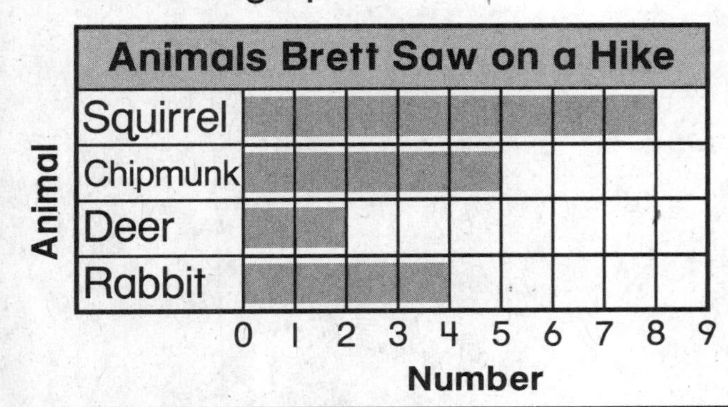

2. How many more squirrels than rabbits did Brett see on the hike?

_______ more squirrels

Measure the length of the real object. Choose the better unit to use.

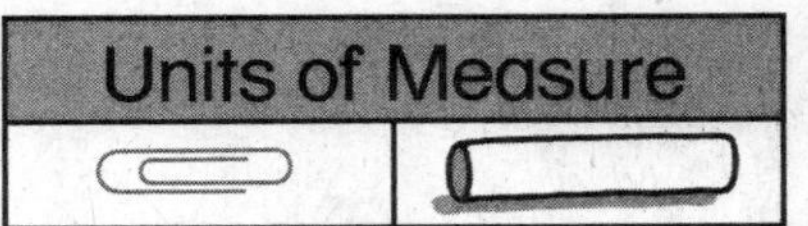

	Find the object.	Choose the unit.	Measure.
3.	bulletin board	____________	about ________

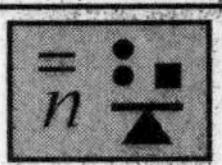

Find the rule. Complete the table.

4. Rule: Add _____.

In	Out
2	5
4	
6	9
8	

5. Rule: Add _____.

In	Out
3	
4	
5	11
6	12

Name ____________________

Time Before the Hour

You can tell the time by saying the number of minutes before the next hour.

Think: the next hour after 1 o'clock is 2 o'clock.

The clocks show 1:45.

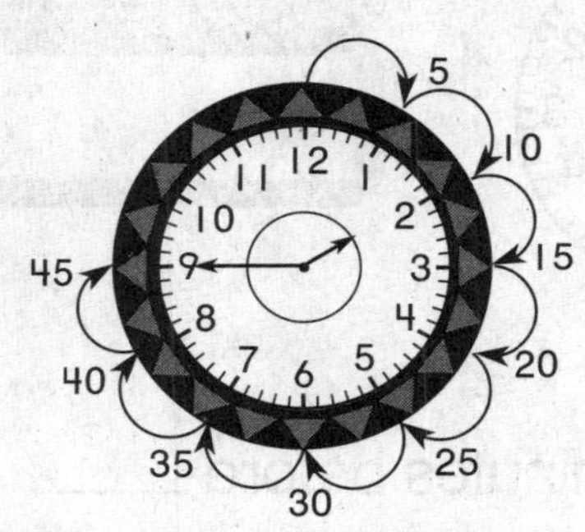

Forty-five minutes have passed since 1 o'clock. It is 45 minutes after 1 o'clock

There are 15 minutes left before 2 o'clock. It is 15 minutes before 2 o'clock.

Write the time in two ways.

1. 50 minutes after 6

____ minutes before ____

2. 40 minutes after 3

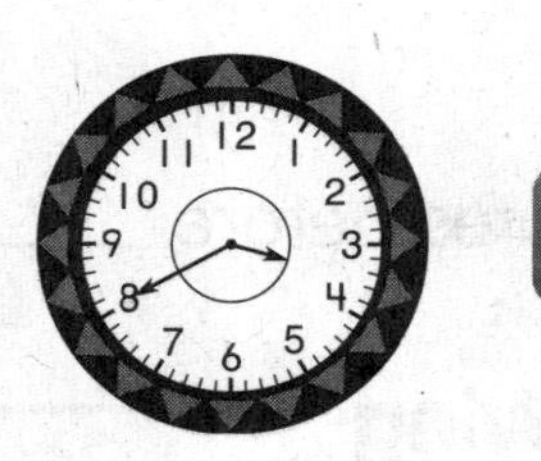

____ minutes before ____

3. 35 minutes after 8

____ minutes before ____

4. 55 minutes after 12

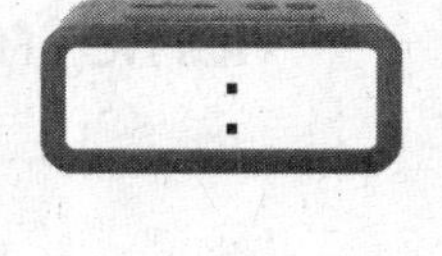

____ minutes before ____

MG1.4 – Tell time to the nearest quarter hour and know relationships of time (e.g., minutes in an hour, days in a month, weeks in a year).

Name ______________________________

Time Before the Hour

Write the time in two ways.

1. 45 minutes after 6

_____ minutes before _____

2. 50 minutes after 2

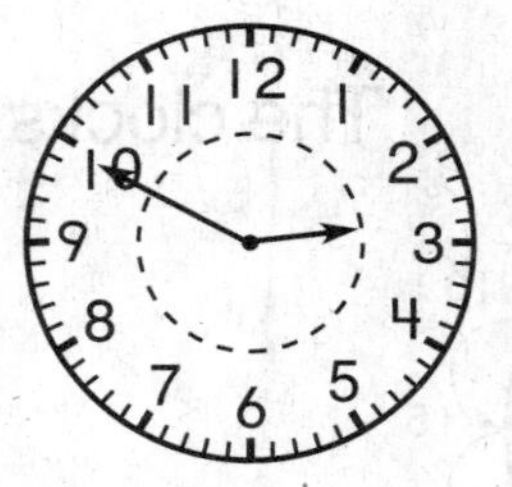

_____ minutes before _____

3. 35 minutes after 9

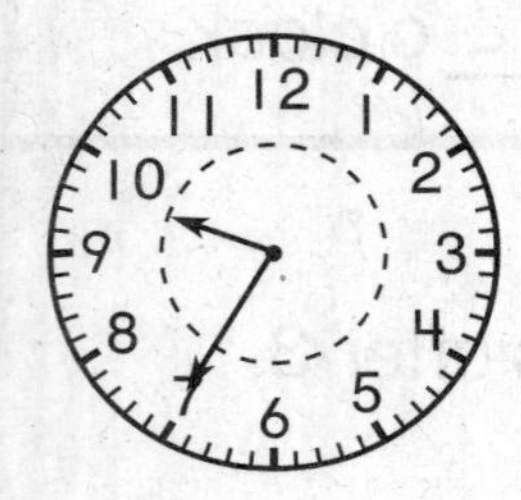

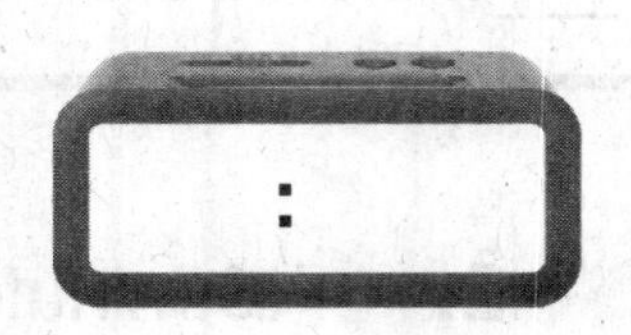

_____ minutes before _____

4. 55 minutes after 5

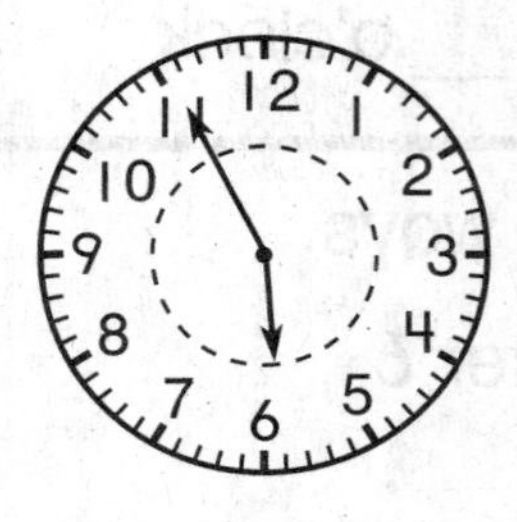

_____ minutes before _____

Problem Solving

5. Deb gets home from school at 45 minutes after 3. Jack gets home from school at 10 minutes before 4. Which classmate gets home from school first?

____

Draw hands on the clocks to show the times.

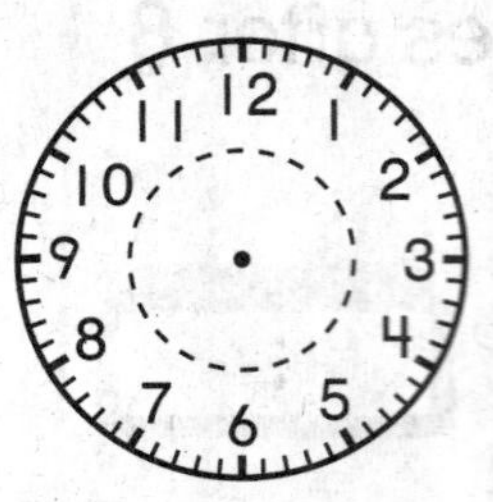

Deb gets home from school.

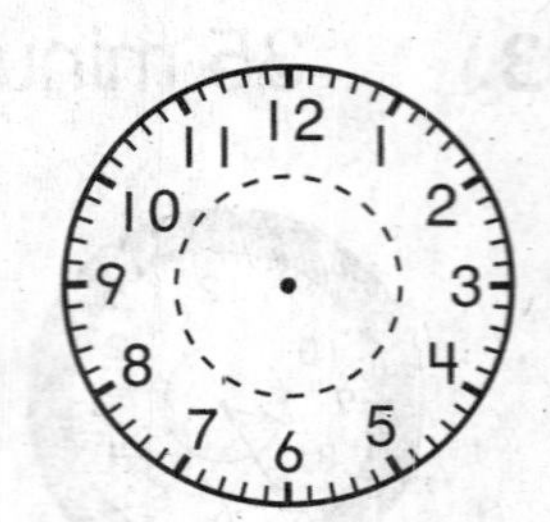

Jack gets home from school.